Groundwater Contamination

GROUNDWATER CONTAMINATION

Sources, Control, and Preventive Measures

CHESTER D. RAIL

Environmental Compliance Specialist, City of Albuquerque
Albuquerque, New Mexico

TECHNOMIC
PUBLISHING CO., INC.

LANCASTER · BASEL

Published in the Western Hemisphere by
Technomic Publishing Company, Inc.
851 New Holland Avenue
Box 3535
Lancaster, Pennsylvania 17604 U.S.A.

Distributed in the Rest of the World by
Technomic Publishing AG

Printed in the United States of America
10 9 8 7 6 5 4 3 2

Main entry under title:
  Groundwater Contamination: Sources, Control,
    and Preventive Measures

A Technomic Publishing Company book
Bibliography: p. 123
Includes index p. 133

Library of Congress Card No. 88-51817
ISBN No. 87762-594-8

To my family

TABLE OF CONTENTS

This book is designed to provide a comprehensive, concise discussion and review of the presently known sources of groundwater contamination and its many complex interactions, including managerial and political implications. Sufficient background information and references on groundwater contamination are presented throughout the text to provide readers with an understanding of the various concepts under discussion. Groundwater contamination interactions are discussed in depth.

Groundwater personnel, specialists, scientists, water resource managers, regulators, water-supply operators, and water-pollution control individuals concerned with activities, evaluations, and preventive measures to groundwater contamination will find this consolidated information important. Other professional groundwater related staff from various disciplines, including personnel from governmental agencies, municipal water-supply and wastewater systems, public health departments, and environmental health entities will also find the information presented in this book valuable. This book, however, is also intended for readers and groups that are interested in and involved with natural resources and concerned about abating and controlling groundwater contamination.

The book can be used as a basic or supplemental text in undergraduate or graduate courses in groundwater contamination, groundwater pollution control, aquatic ecology, or groundwater quality protection and enhancement. It can also be consulted as an environmental reference text by school, municipal, and water resource libraries.

The term *GROUNDWATER* is used as one word (i.e., *groundwater,* not *ground-water* or *ground water*) throughout the text in recognition of its importance as a *single* concept and phenomenon within a complex series of environmental and social forces, all on which we depend. The word *CONTAMINATION* is meant to be analogous to *pollution* throughout, and does not necessarily imply infectious organisms or a state thereof.

ACKNOWLEDGEMENTS

I would like to acknowledge the numerous contributions made in the area of *Groundwater Contamination* by the U.S. Geological Survey and the U.S. Environmental Protection Agency. Truly, they have conducted and coordinated many substantial studies and reviews dealing with this subject area and are the real experts. For preparation of this manuscript I am also grateful for the assistance rendered to me by the staff of the University of New Mexico, Zimmerman Library, Office of Government Documents, Federal Regional Depository. Without their assistance, many of the documents referenced in this book would not have been available for review. And, of course without the help of my loving wife Carole Jane, and two sons Brandon and Jason, this book would not have been completed.

Groundwater as a Resource

Groundwater constitutes a significant source of water in most areas of the United States and may be defined as subsurface water that occurs beneath the water table in soils and geologic forms that are fully saturated (Freeze and Cherry, 1979; Pye et al., 1983). Groundwater is also an integral part of the hydrologic cycle, an important natural resource, and any evaluation of contamination problems should recognize this (Van der Leeden, 1971). The hydrologic cycle, according to the U.S. Geological Survey (1978), is a natural machine, a constantly running distillation and pumping system, which is supplied by energy from the sun (Figure I). Gravity keeps water moving from the earth to the atmosphere via the vehicles of evaporation and transpiration. Water moves from the atmosphere to the earth in the form of condensation and precipitation. It then moves between points on the earth in streamflow and groundwater movement systems. This process of water movement throughout the earth has neither a beginning nor an end, with oceans constituting the major source of water, the atmosphere as the deliverer, and the land being the principle consumer. This closed ecosystem does not lose water or gain it, but the amount of water available to eventual users fluctuates because of variations which can occur at the source. The U.S. Geological Survey (1978) summarizes aspects of the hydrologic cycle as follows:

- Every second, millions of molecules heated by the sun evaporate and supply water for the hydrologic cycle.
- The water supply of the world is estimated at 326 million cubic miles (i.e., more than 326 trillion gallons) with only about 5 gallons of every 100,000 gallons in motion (i.e., most of the water is stored in the oceans, glaciers, lakes, or underground).

- The United States receives approximately 1,430 trillion gallons of precipitation each year.
- Only about 3,100 trillion gallons of water in the form of invisible vapor occur in the atmosphere (if it were to fall at once, the earth would be covered with about one inch of water).
- Water, once fallen, however, regardless of how long it may be delayed or held (e.g., in glaciers, lakes, etc.) is eventually released to enter the cycle once more.
- Of the 102,000 cubic miles of water passing into the atmosphere annually, 78,000 cubic miles fall back into the oceans, with streams, rivers, and groundwater also collecting and returning to the ocean 9,000 cubic miles of water. The remaining 15,000 cubic miles of water maintain life processes, principally as soil moisture, which provides water necessary for vegetation, although this water, too, reaches the atmosphere again by the process of evapotranspiration.
- More than 2 million cubic miles of fresh water is stored in the earth in the form of groundwater, half in upper surface areas, with more than 35 times the amount held on the surface in lakes, rivers, and inland seas. This, however, is still relatively small compared to the 7 million cubic miles stored in glaciers and icecaps.
- The 317 million cubic miles of water (i.e., more than 317 trillion gallons) held by the oceans constitutes 97.3 percent of the earth's supply.

The ability of an aquifer (i.e., a water bearing stratum of permeable rock, sand, or gravel) to store and transmit water is a function of its porosity (i.e., having pores that permit liquids to pass through) and permeability (i.e., to liquids)

Introduction

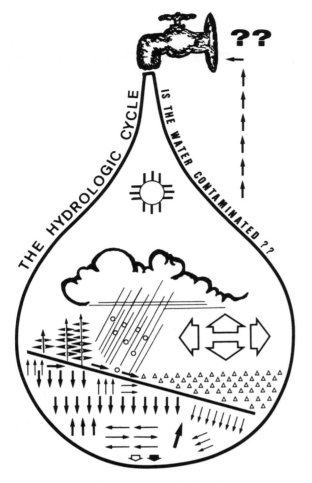

Figure I. The hydrologic cycle.

(Freeze and Cherry, 1979; U.S. EPA, 1977; Pye et al., 1983). However, little is still known about the characteristics of aquifers in many areas of the United States (Conservation Foundation, 1985). Information is needed about physical configurations of aquifers, including boundaries and linkages between other bearing areas, chemical properties, sensitivity of recharge and discharge areas, location of current wellfields, rates of recharge and discharge, and location of pockets of highly saline or contaminated water. Investigations now should focus on aquifers that are current or potential drinking water supply sources. Emphasis should be placed on fields containing present or future planned drinking water supply wells (National Research Council, 1977, 1978, 1979, 1980, 1981, 1982, 1983). Federal, state, and local governments should collect detailed information on current uses made of their groundwater resources and investigate potential sources of contamination.

Localities in the United States, specifically in the northeastern states, have developed aggressive enforcement programs that require an administrative order process to resolve groundwater contamination problems (Savage,

1986). Contaminated municipalities are being obliged to perform diagnostic studies, do site cleanup work, or provide an alternative water supply. Accordingly, significant preventive programs also include:

- a formal use classification system to identify existing and potential sources of drinking water and establish withdrawal policies
- expansion of regulatory authority to underground fuel storage, septic tank system additives, septic tank system design, and septic tank installation
- institution of permit programs for groundwater withdrawals
- initiation of strategies to improve protection of groundwater through public education and development of municipal groundwater protection programs

In western and southern states communities have also adopted groundwater quality standards along with comprehensive classification systems (Savage, 1986). In many instances, water is classified where there is a threat to contamination, and many management plans are being designed to protect human and environmental health by preventing and mitigating groundwater contamination.

The land use decision-making process is a key element in how groundwater resources will be protected in the future. Efforts must be made to work with municipalities to ensure that future land use is consistent with long-term protection of groundwater (see Checklist in Appendix B).

Groundwater Movement

Groundwater moves in response to gravity, pressure, and friction. The first two drive the water, and the latter restricts its motion (Pye et al., 1983). Due to the complexity of groundwater flow, it is difficult to construct models of movement at the microscopic level. The French engineer, Darcy, however, formulated an empirical law which averages microscopic complexities and provides a model for groundwater movement. Darcy's Law relates groundwater flow across a surface to the rate of change of energy along its flow path (Pye et al., 1983; Todd, 1960).

Groundwater flow and contamination transport models have been widely used for planning and design purposes in the past decade (Wen-Sen-Chu et al., 1987). Additionally, concerns about groundwater contamination have motivated the development of many computer simulation models for groundwater quality management (Remson et al., 1971; Bachmat et al., 1980; Javandel et al., 1984). These models characterize groundwater flow and contaminant transport with varying levels of simplification.

Also, the main difference between groundwater and surface flow is that movement of groundwater is slower. Flow rates of groundwater are governed by hydraulic gradients and aquifer permeabilities which can range from a few millimeters to several meters per day. Movements of groundwater generally occur from a point of recharge to a point of discharge along a hydraulic head. Usually the laminar flow which occurs results in little mixing of the groundwater, when compared to that which occurs when surface water moves. In fact, in groundwater environments, the rates of contaminant movement are determined by groundwater flow-rates, physical interaction between aquifer materials and contaminants, and changes in water chemistry (Pye et al., 1983).

The chemical composition of natural groundwater can also be derived from many different sources, including gases and aerosols from the atmosphere, weathering and erosion of rocks and soils, solution or precipitation reactions occurring below the land surface, and cultural effects resulting from activities of humans (Hem, 1959, 1970). Precipitation of minerals in groundwater can be closely evaluated by studying chemical equilibrium functions, including the law of mass action and the Nerst equation (Hem, 1970). Contaminated groundwater, basically, cannot be diluted by the entire body of related groundwater, as is the case with surface water. Little dilution of groundwater takes place in fractured rock or carbonate aquifers (Pye et al., 1983), and in unconsolidated deposits, dilution of groundwater occurs primarily by dispersion (Miller, 1981).

Groundwater Usage

A most significant trend in the water-use picture is the changes which have occurred since colonial times (UNESCO, 1980). In the colonial period, water was considered to be an essential free resource and now it has become an expensive commodity in many locations (U.S. Geological Survey, 1980).

English common law regarded water as a common property resource. Those who owned land along streams could use it freely. This common law doctrine was brought by the colonists to the eastern United States where it still strongly influences water laws of each appropriate eastern state. Water legislation, however, in most of the western United States was influenced by Spanish law and customs. It was generally adapted to meet the needs of farmers, miners, and ranchers in arid regions. The doctrine which evolved as an element of western water law held that the person who first diverted water and put it to a beneficial use had a right to maintain diversion and use. Subsequent upstream diverters, consequently, could operate only if prior water rights were being satisfied.

As water needs increased for fire protection, dust control, sanitation, disease eradication, and domestic consumption, an increase in the extent of water supply systems occurred. These water uses could be grouped into three categories (James et al., 1980):

(1) "Withdrawal uses," such as withdrawal from a well or diversion from a stream for public supplies, irrigation, and industry

(2) "Non-withdrawal uses," such as for hydroelectric power, navigation, recreation, preservation of wildlife and sport fishing habitat, salinity control, waste dilution, and transport

(3) "Non-supply uses," such an evaporation from lakes and reservoirs and evapotranspiration from non-irrigated food and fiber crops

Past and present water uses can be estimated, but total water needs are changing. At present, water use is responsive to prices, technology, customs, and regulations. Although inventory systems for estimating water use can indicate past and present uses, such predictions are not accurate or frequently reported. Future water use trends are also often not worked out (James et al., 1980).

The average per capita water consumption in 1983 was approximately 159 gallons per day (Pye et al., 1983; Last, 1980). In some areas it exceeds 200 gallons per day. Exterior residential use may range from 5 to more than 150 percent of the interior use, averaging about 75 percent. This water use is primarily for watering of lawns and gardens. Personal use of water (e.g., laundry, toilet flushing, etc.) accounts for 40 percent of internal residential use and averaged 6.3 gallons per capita per day in 1983.

Safe Yield

The safe yield of a groundwater basin is defined as "the amount of water withdrawn annually without producing undesired effects" (Todd, 1959). Undesired effects include: depletion of groundwater reserves, intrusion of water of an undesirable quality, contravention of water rights, the deterioration of the economic advantages of pumping, excessive depletion of stream flow, and land subsidence (Pye et al., 1983; Domencio, 1972; Freeze and Cherry, 1979). Excess withdrawal of a safe yield is termed *overdrafting*. Groundwater overdrafting is considered serious if it continues indefinitely (Pye et al., 1983; U.S. Government Accounting Office, 1980).

Providing complete and accurate water use information will involve input from hydrologists, engineers, and economists. It will require detailed and complete water use data and knowledge of factors related to groundwater "quality" and "quantity" concerns. Water use information must be collected by federal, state, and local agencies, in a properly planned, systematic effort to summarize reliable data to serve a multitude of purposes.

This period of free and easily developed water supplies has ended. In fact, water use now exceeds available supply (James et al., 1980). We are not running out of water, but beginning to recognize that many different types of groundwater contamination problems exist and they should be corrected. That is what this book addresses. It identifies and describes "sources of contamination in groundwater" and proposes alternatives and options for control or abatement of groundwater problems. Its central theme is that *prevention of contamination problems in groundwater is the best solution.*

Groundwater Contamination

Sources and Control

Groundwater flows in subsurface areas of the earth. Groundwater in its many forms carries nourishment to human individuals, wild animals, and into and out of ecological communities. Groundwater sources also support economic activities, with some involving food, energy, mining and industrial concerns. Underground water areas that exist are replenished by surface water, precipitation, and direct discharges. The quality of groundwater continues to be threatened by natural and human activities, although contamination of underground aquifers usually is a consequence of people's misuse of the environment (UNESCO, 1980).

Presently in the United States, the national extent of the threat to groundwater is receiving much attention. This is primarily because of the many thousands of publicized well closures and contamination incidents which have occurred throughout the country. In addition, over 200 organic and inorganic chemicals in excessive amounts, including high levels of radioactive isotopes, have been identified in some groundwater supply systems. Attention has been given to these areas because of the direct threat to public health of the individuals that consume the water. Increased costs associated with location of new water supply systems (i.e., if new sources of supply have to be located), pumping water from remote locations or deeper areas, having to provide increased levels of preliminary and final treatment to meet standards related to safe consumption of the water, have caused a financial crisis in many areas. Groundwater contamination and its control are not only technically complex but involve numerous economic and political demands.

Groundwater Contamination Problems

The quantity of groundwater which underlies North America is truly immense and the amount present is six times greater than the water stored within surface reservoirs (Conservation Foundation, 1985). People depend upon groundwater in every state, and its usage accounts for approximately one-fourth of all water used. This consumption includes the following functions and statistics:

- Thirty-five percent of water withdrawn for municipal water supplies comes from underground sources.
- Groundwater as a sole source of drinking water is

used by more than 50 percent of the total United States population and 97 percent of rural residents.

- Forty percent is agricultural irrigation water usage and 26 percent is industrial withdrawals (excluding electric power plants).
- Nourishment and maintenance of ecosystems is valued for fish production, wildlife habitat, and recreational opportunities.

Generally, most groundwater supplies in the United States are of good quality and produce essential quantities. The full magnitude of groundwater contamination in the United States is, however, not fully documented and federal, state, and local efforts continue to assess and address the problems (U.S. Geological Survey Yearbook, 1980; Rail, 1985a). The nation is not facing an overall groundwater contamination crisis, but in some localities, local contamination has caused significant well closures, public health concerns, and economic losses (Burmaster and Harris, 1982). Health concerns aroused by the large number of contamination cases discovered during recent years and systematic studies presently underway reveal the specter of a more widespread problem. Additionally, local and aggregate impacts of groundwater contamination discoveries evoke greater recognition of the important role of good quality groundwater. Contaminated groundwater has limited utility and deterioration of its quality constitutes a permanent loss. This is because treatment or rehabilitation of an aquifer, in many instances, is impractical and/or economically unfeasible.

One-half of the United States residents rely on groundwater which is not treated or disinfected as their source of drinking water (Council of Environmental Quality, 1981). Apparently, over several decades, human activities have contaminated drinking water supply wells with "traditional" contaminants. For example, in sections of the midwest, the use of agricultural fertilizers raised nitrate concentrations in groundwaters above federal safe drinking water standards. In the snow belt, also, hundreds of wells were closed because of high concentrations of salt percolating into groundwater from de-icing agents used to maintain travel on highways during winter months. Groundwater in every section of the country (i.e., private and public areas) has also been contaminated over the years by bacteria and viruses, often as a result of improperly designed or maintained on-site liquid waste disposal systems (e.g., septic tanks, leachfields, leaky private or municipal sewer pipes).

Additionally, in the past few years, newspapers, national magazines, and TV news media have carried stories and reports concerning groundwater contamination problems throughout the country. The incidences reported involved discoveries of high concentrations of organic chemicals in municipal community or non-community private drinking water supply systems. As a result, thousands of drinking water supply wells, which provided water for millions of individuals, were closed. Excessive amounts of toxic chemicals that are considered suspected carcinogens were present. It was also impossible to say with any degree of certainty how long specific contaminants had existed. Determining the full exposure that animals and/or humans had received in these areas via drinking water was also difficult to quantify. In fact, more questions and uncertainties were brought forward because of different types of groundwater contamination. The magnitude and distribution of groundwater contamination is not always understood.

General Groundwater Contamination Sources

Contamination of groundwater from human activities and natural situations includes many sources (Figure II). Sources in this instance are defined as points along a pathway, originating in human activities, through which substances travel as they flow through and are released into groundwater (Office of Technology Assessment, 1984). Substances are stored in, or flow through, sources in a variety of ways. They are derived from the storage of raw materials (e.g., product storage), distribution (e.g., pipelines), storage (e.g., underground storage tanks), use (e.g., pesticide applications, etc.), and actual disposal (which can take place anywhere in the previously mentioned processes).

A study conducted by the Office of Technology Assess-

Figure II. Sources of contamination.

ment (OTA, 1984) recognized thirty-three sources known to contaminate groundwater. The OTA study categorized these sources according to the nature of their substance release to groundwater. The six categories established by OTA included:

(1) Sources designed to discharge substances
(2) Sources designed to store, treat, and/or dispose of substances (i.e., discharge occurring through unplanned release)
(3) Sources designed to retain substances during transport or transmission
(4) Sources discharging substances as a consequence of other planned activities
(5) Sources providing a conduit discharge through altered flow patterns
(6) Sources (e.g., naturally occurring) whose discharge is caused solely by human activity

Other categorization schemes of groundwater contamination sources are also possible (e.g., according to the nature of the user: agricultural, industrial, domestic, municipal or the physical location of the source; above the land surface, below the land surface and above the groundwater table, and below the groundwater table) (OTA, 1984). Classifications based on discharge systems, however, have an advantage of identifying the entry of substances into the environment.

The OTA (1984) report also mentioned that "points-of-entry" are places where evaluations can be conducted initially to discover and alter entry (e.g., detection, evaluation, correction, and prevention of contamination) and that three conclusions can be reached based on the previously presented classification scheme. The three conclusions include:

(1) There is a great diversity of sources associated with a broad range of industrial, agricultural, commercial, and domestic activities (i.e., wastes and non-wastes are potential contaminants of groundwater) with most attention focused on hazardous wastes from a point source (i.e., an easily identified facility, such as a landfill or impoundment).
(2) Only a few types of discharge facilities are specifically designed to accommodate substances released into the subsurface.
(3) Non-waste releases can result from sources designed to retain non-waste products (i.e., discharges occurring through an unplanned release).

The occurrence of substances in groundwater and an understanding of how, why, and where they are present, also (according to OTA, 1984), are directly related to their categorized use and/or disposition. The categories listed include:

- *Category I*—subsurface percolation, injection wells,

land application of wastewater, wastewater by-products, hazardous wastes

- *Category II*—landfills, open dumps, residential disposal, surface impoundments, waste tailings, waste piles, materials stockpiles, graveyards, animal burial, aboveground storage tanks, underground storage tanks, containers, open burning, detonation sites, radioactive disposal sites
- *Category III*—pipelines, materials transport, transfer operations
- *Category IV*—irrigation practices, pesticide applications, fertilizer applications, animal feeding operations, de-icing salt applications, urban runoff, percolation of atmospheric pollutants, mining and mine drainage
- *Category V*—production wells, including oil, geothermal and heat recovery, water supply, and other wells, plus construction excavation
- *Category VI*—groundwater–surface water interactions, natural leaching and saltwater intrusion

Nine groupings were also associated by OTA with the previously listed sources of groundwater contamination (Woodward-Clyde consultants, 1983; from OTA, 1984). These groupings included:

Organic chemicals
(1) Aromatic hydrocarbons
(2) Oxygenated hydrocarbons
(3) Hydrocarbons with specific elements
(4) Other hydrocarbons

Inorganic chemicals
(5) Metals/cations
(6) Nonmetals/anions
(7) Inorganic acids

Other
(8) Biologicals
(9) Radionuclides

The conclusions that were reached by OTA from the previously listed categories and groupings included:

- that certain substances are associated with known sources of contamination, with metals/cations and non-metals/anions being common, followed by hydrocarbons (e.g., pesticides, chlorinated solvents), miscellaneous hydrocarbons (e.g., fuels), and radionuclides
- that the association of substances with specific sources of contamination will differ based on disposal patterns of different segments of society (i.e., pesticides may enter groundwater from storage tanks, from aerial spraying during agricultural oper-

ations, and from residential disposal into backyards or sanitary landfills; radioactive disposal sites, on the other extreme may not contain organic chemicals; consequently, generalizations about the association of substances with sources are not possible in most instances)
- that known contaminated areas are likely to release a variety of chemical substances, with each one showing different chemical properties, including a significant variability in the toxicity of products produced
- that in most instances, a broad range of toxic substances may be present in groundwater, with the presence of each chemical being dependent on past and present land uses

Identification of Groundwater Contamination Risks

The reliable assessment of risks arising from groundwater contamination problems and the design of efficient and effective techniques to mitigate them, require the capability to predict the behavior of chemical contaminants in flowing groundwater (National Research Council, 1984). The National Research Council (1984) supported the concept that: "Reliable and quantitative predictions of contaminant movement can be made only if we understand the processes controlling transport, hydrodynamic dispersion, and chemical, physical, and biological reactions that affect soluble concentration in the ground."

In addition, the widespread use of chemical products by society, coupled with the disposal of a larger and ever increasing volume of waste materials, poses the potential for widely distributed groundwater contamination, with new instances of groundwater contamination by hazardous chemicals (e.g., pesticides, herbicides, and solvents) continually being recognized by the NRC (1984). This is because chemicals which are used ubiquitously in everyday life now have achieved widespread use in urban, industrial, and agricultural settings and eventually reach groundwater and contaminate it. Such contamination occurrences, in many instances, then pose direct serious threats to public health concerns. In addition, when groundwater areas are contaminated, they usually remain so for decades because of the slow rate of groundwater movement and the natural flushing of aquifers which occurs in most underground storage areas and geological formations.

The challenges set forth by the National Research Council (1984) to abate groundwater contamination basically include:

- to ultimately prevent the introduction of contaminants into an aquifer
- to predict contaminants' movement, once they are introduced

- to remove them, to the extent possible, in ways by which the biosphere can be protected effectively

The council (NRC, 1984) also maintained that the largest potential source of contamination of groundwater involves the improper disposal of solid and liquid wastes. This is because during the past several decades, legislation reflecting environmental concerns restricting air and surface water pollution resulted in increased disposal of wastes into subsurface areas. For example, in 1980 the United States Environmental Protection Agency (U.S. EPA, 1980) estimated that 200,000 landfill sites and dumping areas received 150 million tons per year of municipal solid wastes and 240 million tons of industrial solid wastes. Surface impoundments (i.e., in addition to landfills) received 10 trillion gallons per year of liquid industrial wastes. The U.S. EPA (1980) estimated that 142,000 tons of hazardous wastes were generated daily in the United States (e.g., 60 million tons annually) at more than 750,000 sites. They also estimated that 50,000 larger industrial sites, in the past, had landfills on their property, and such areas were used for the disposal of hazardous wastes. In the same report (U.S. EPA, 1980), septic tank systems, agriculture, accidental leaks and spills, mining, highway de-icing, artificial recharge, underground injection, and saltwater encroachment, were also listed as additional sources that cause groundwater contamination in the United States.

Current estimates by the National Research Council (1984) on the extent of groundwater aquifer contamination in the United States suggest that 0.5 to 2.0 percent may be polluted. Although overall, this might not appear significant, known contaminated areas are highly dependent on groundwater as their principal and sole means of water supply.

Subsurface Disposal of Wastes

The disposal of wastes in subsurface areas has always been viewed as attractive because of the slow movement of water in comparison to the fast movement of surface waters. The thought of "out of sight—out of mind" appeared to be sufficient to solve problems of disposal of certain wastes, whether they be solid and/or liquid in nature. Unfortunately, as we know today, this was not always the proper thing to do because the cost associated with cleanup or decontamination of *some* of these sites (i.e., it is recognized that not all releases of waste material or products cause detrimental effects on groundwater) in later years has proved to be expensive. The costs do not only occur in the form of financial commitments necessary to decontaminate areas, but in the increase of diseases that often are traced to original waste disposal products or toxic by-products. Payment for decontamination or for treatment of diseases will be made for disposing of toxic waste products within subsurface

areas and the question of "Should society pay now or later?" is certain because payment will be made.

United States federal environmental legislation (Resource Conservation and Recovery Act, Water Pollution Control Act, Safe Drinking Water Act, the Comprehensive Environmental Response Compensation and Liability Act) have further restricted ground disposal of wastes. The legislation which restricted the use of air and surface water for disposal of wastes now limits the use of land for waste disposal. So what can be done? Modern and civilized societies must dispose of various large quantities of hazardous or toxic wastes that are generated without increasing groundwater contamination problems. Toxic substances should be recycled, neutralized, and/or stored in properly constructed repository sites. What then is the problem? Is it that perhaps as a society the United States has failed to move towards developing a rational and workable strategy that provides long-term protection to individuals and the environment (NRC, 1984)? Are we too greedy a nation? Must value systems change to allow society to act responsibly within the best technical knowledge available and not only when crisis situations are brought to the forefront? Will our nation act responsibly? What must be done? Answers to some of these questions will be discussed throughout this book. However, many more questions will remain unanswered.

Cleanup Costs

Cleanup of contaminated aquifers to acceptable limits is an expensive and time-consuming process. The restoration involves drilling many wells and pumping vast quantities of groundwater for cleanup purposes, with the pumped water (i.e., after it is treated properly to reduce or eliminate contaminants) reinjected into the aquifer or deposited into a different place. The estimated costs for containment of groundwater contaminants at the Rocky Mountain Arsenal, Colorado, for example, were approximately $100 million in 1982, with cost estimates for "total decontamination" of the arsenal ranging from $800 million to $1 billion (U.S. Army, 1982; from NRC, 1984). The cost/benefit trade-offs to society in these cases need to be examined carefully. In all probability, it should be recognized that some sites are so expensive to restore they may have to be designated as "permanently contaminated." However, if such areas are to be maintained as repositories, they should be selected, designed, and engineered in terms of hydrology, geology, hydrogeochemistry, and microbiology of that particular site. The characteristics of the specific wastes that have been identified to occur or to be deposited in any particular repository in question should also be evaluated (NRC, 1984). The process of selecting adequate repositories is complex, because extreme care must be taken to provide isolation of potential or actual toxic substances from the biosphere for long periods of time. High-level radioactive wastes must be isolated from the biosphere for thousands of years (1,000–10,000 years +), and other nondegradable highly toxic substances also require an organized attempt for permanent isolation.

Scientific knowledge and engineering techniques and modifications must be brought to the forefront and applied to the evaluation of groundwater contamination (NRC, 1984). Yesterday's mistakes should not be repeated, although complex groundwater contamination problems always continue to evade the best knowledge. Substantial and significant efforts must also be made to understand physical-chemical-biological interactions and the overall effects of groundwater contamination on ecosystems. If adequate concentrated efforts to evaluate groundwater contamination problems occur and priorities of evaluation are kept in pespective, toxic substances and hazardous waste in subsurface areas in different ecological environments would not pose short- nor long-term hazards.

ONE

Historical Aspects of Groundwater Use

Early Use of Groundwater for Water Supply

Human settlements normally began at sites where a supply of water for drinking and other necessary purposes was available on the surface, generally from springs or streams (Bromehead, 1942). Settlements once founded, grew, expanded, and eventually outgrew their original water supply or contaminated it by their very existence and by depositing their waste materials into it. In other situations (i.e., location of strategic sites for wartime protection or location of valuable minerals or commodities; Clark, 1944) settlement sites were established without direct regard for availability of water.

Primitive man paid little attention to water in humid areas because it was always present. Like air, it was taken for granted (Tolman, 1937). However, in semi-arid and arid regions of the earth, the occurrence of water did control activities of human beings and their settlements. Villages and tribal units were constructed adjacent to flowing streams, water holes, and springs. Primitive man's migrations also had to follow perennial water courses and this was essential, especially during drought.

Ancient man also knew how to excavate for water. This might have been learned by observing wild animals in their search for water (Tolman, 1937). However, regardless of how the local well was discovered, it soon became an important possession, especially if domesticated animals such as cattle or sheep were kept.

Bromehead (1942) classified the methods adapted by man to obtain water into two main groups. These two groups included "wells" and "aqueducts." The word *well* as defined by Bromehead meant: "a means of obtaining water from the earth vertically beneath the spot at which it is required, where it is not obviously present at the surface." The word *aqueduct* as defined by Bromehead signified: "when water is visible at the surface but not at the desired spot it may be brought to it by pipes or artificial channels, that is by *aqueducts* in the literal sense of the word."

Springs have also remained foci of human settlements and are capable of yielding valuable information to the discerning archeologist (Clark, 1944). In fact, a common device for securing a ready supply of water from a natural spring was to reinforce the sides with stones or wood. A sufficient depth of water could then be accumulated and water could be collected from it with buckets, dippers, or other receptacles. Enclosed springs are considered the first improvement on natural springs and are traceable to remote antiquity (Clark, 1944).

Wells and Springs as Water Supply Systems

The first well, utilizing strictly scientific principles to reach water, appears to have been sunk in Derbyshire in 1795, although qanats (kanats, ganats) were invented in Iran several thousand years ago, and some wells were dug in China during that same time period (Wulff, 1968; Bromehead, 1942). Qanats will be discussed in more depth in this chapter.

Clark (1944) emphasized that as the depth of wells increased it made it necessary to cover well heads (if only to prevent children from falling into the shafts). Also, in most instances, wells, once constructed, had to be maintained. As a result of this, wells had various forms of steps or foot-rests associated with their shafts. The day-to-day operation of wells was burdensome. It involved hauling water up the shaft (i.e., although in many instances it was lightened by a variety of ingenious lifting appliances). Some of these lifting appliances are described by Tolman (1937) and Smith (1975).

Wells and their importance can be traced in the Book of Genesis (it is reported that Abraham dug wells, some of which were later filled up by the Philistines and redug by

Jacob). Wells were so valuable that individual names were given to them as an indication of their worth.

The Bible recounts numerous incidents which illustrate the importance of groundwater supplies to the Tribes of Israel. Abraham and Isaac were known for their dug wells and as stated by Meinzer (1934), "the twenty-sixth chapter of Genesis . . . reads like a water supply paper." Early romances (Rebekah, Genesis 24; Rachael, Genesis 29) were also enacted at well sites (Tolman, 1937).

Jacob's well is one of the few water supply systems mentioned in the Bible that is identified by many authorities (Tolman, 1937). The well is a rectangular enclosure (151 by 192 feet) surrounded by walls which have crumbled to ruins. The well in 1937 was 75 feet deep and 7½ feet in diameter. The upper portion of the well was lined with rough masonry and filled up to half its depth by fallen debris (Schaff, 1885). During their wanderings, the Israelites also dug wells under the direction of the Lord who promised them "houses full of all good things, which thou fillest not, and wells digged, which thou diggest not" (Deuteronomy 6:11).

Miracle-working waters are also described in the Christian Bible. In John 5:2–9, we read:

> Now there is at Jerusalem by the sheep market, a pool, which is called in the Hebrew tongue Bethesda, having five porches.
> In these lay a great multitude or impotent fold of blind, halt, withered, waiting for the moving of the water.
> For an angel went down at a certain season into the pool, and troubled the water: whosoever then first after the troubling of the water stepped in, was made whole of whatsoever disease he had.

According to Tolman (1937) the pool of Shiloah where Christ healed a blind man was in use for many years and had a "good supply of water, generally somewhat salty to the taste, perhaps from the soil through which it percolates, and it is, moreover, polluted by the washerwomen and tanners by whom it is constantly used. . . ." The water according to Tolman (1937) is supplied from the Fountain of the Virgin, a tunnel 1,708 feet long cut into solid rock.

Roman Groundwater Development

Archeologists relate that before the building of the first aqueduct in 312 B.C., the Romans depended for their water supply upon the Timber and upon wells, springs, and rainwater caught and stored in cisterns (Tolman, 1937). The soil at that time was rich in springs and underground streams, and wells could be sunk successfully at any point with the average depth being 5 meters. Such wells were common from these periods. In fact, excavations from the area known as "The Forum" have uncovered more than 30 wells, some of which date from the Roman Republic (Platner, 1911; from Tolman, 1937). However, according to Tolman (1937), the near-surface groundwater of the Pontine marshes was so contaminated that the Romans abandoned their wells and

imported water by means of aqueducts. They never utilized the groundwater supplies of their far-flung empire. Apparently, the Romans were prejudiced against the use of groundwater.

Butler (1902) in discussing ancient water wells in Athens around the middle of the sixth century B.C. stated: "Many of the houses are provided with deep wells stoned up with polygonal masonry and nearly all have conveying cisterns and galleries. . . . In the wider streets are public wells of great depth, covered by stone slabs, with small apertures, the necks of which are well furrowed by the ropes which for centuries have drawn the dripping buckets from their cool depths. Some of them are again in use." The wells in Athens, as in other areas also seems to have always been the center of rural urban life and of apparent feminine activity and gossip (Tolman, 1937).

Artesian Wells in France

In Artois and Modena (northern Italy), in 1126, popular and scientific interest focused on the occurrence of sprouting water. And here, according to Tolman (1937), medieval settlers developed the art of drilling wells instead of digging them. The development of drilling wells (i.e., as opposed to digging them) has also been described by Norton (1911) and Bowman (1911). They recognized that the art of drilling and casing wells was invented, perfected, and extensively practiced by the ancient Chinese. It seems that the Chinese by using bamboo materials could, with much patience, penetrate great depths into soil formations. In fact, according to Tolman (1937), Professor G. D. Louderback described to him how wells could be drilled to depths of 5,000 feet by the use of bamboo and manual churn drilling. The use of bamboo casing is, of course, also essential. When these wells were drilled, they were started by the grandfather and completed by the grandchildren.

Ancient Persians also constructed tunnels and shafts to tap groundwater sources (Johnson Division, 1975; Wells 1924). Near the end of the 11th Dynasty (about 2100 B.C.), one leader of Mentuhotep's Egyptian forces also reported sinking 14 wells, utilizing an army of 3,000 men (Johnson Division, 1975).

Throughout the ages, humans settled in regions where water was plentiful and of good quality (USDA, 1955). However, when supplies failed or were made useless by unbearable silt, contamination, or when floods eliminated everything, these centers of habitation were abandoned. But often the causes lay as much in the acts and failures of men themselves as in the caprices of nature. Man contaminated water sources to the point where the water was unfit for consumption, although he also constructed elaborate water-supply systems. These ancient wells, aqueducts, and reservoirs, some still serviceable after thousands of years, attest to civilization's capacity for constructive thinking (USDA, 1955).

The people of Assyria, Babylonia, Egypt, Israel, Greece, Rome, and China, built similar water supply facilities long before the Christian era. In fact, Egypt has the world's oldest known dam, a rock-filled structure built 5,000 years ago to store drinking and irrigation water and perhaps also to hold back floodwaters (its length was 355 feet, and its crest 40 feet above the riverbed).

Among the early Greeks, Hippocrates recognized the dangers of contaminated drinking water to health and recommended that the water be filtered and boiled (USDA, 1955). Based on this knowledge, the Romans used their poorer water for irrigation and fountains.

The Tukiangyien systems, used in China some 2,200 years ago, is another tribute to the genius and toil of ancient people. This skillfully designed multi-purpose engineering project was intended to divert flows of the Min River, a stream that rises on the high plateau of Tibet (USDA, 1955). By building dams and dikes on the main river, farmers divided its flow into many parts so they could irrigate one-half million fertile acres.

In H. G. Wells' *Outline of History* (Wells, 1924; Vol. 1) the author states:

> No creature can breathe, can digest its food without water. We talk of breathing air, but what all living things do really is breathe oxygen dissolved in water. The air we ourselves breathe must first be dissolved in the moisture of our lungs: and all our food must be liquified before it can be assimilated. . . . The lower plants are still the prisoner attendants of water. The lower mosses must live in damp, and even the development of the spore of the ferns demands at certain stages extreme wetness. . . . And after the plants came the animal life. There is no sort of land animal in the world, as there is no sort of land plant, whose structure is not primarily that of a water-inhabiting being which has been adapted through the modification and differentiation of species to life out of the water.

Many habits of humans and their social organizations have been influenced more by their association with water than with the land by which they earned their bread (USDA, 1955). This association is reflected in the psalms of the Hebrew poets and in the laws, regulations, and beliefs among the civilizations of the Near East, the Far East, and South America. Many of the permanent settlements in arid regions, today as in antiquity, have concentrated along river valleys and along seacoasts. Human habitated areas have always clustered around or near convenient sources of fresh water. Transportation has also been affected by this.

The Incas at Lake Titicaca in Peru employed rafts cleverly constructed of native plants (USDA, 1955). Solid logs filled with double outriggers and platforms made seaworthy craft in Africa and Polynesia; later craft were constructed from logs hollowed out by fire and crude tools (Golson, 1963; Taylor, 1951; Enoch, 1912; Grieder, 1982).

Inland transportation since early times has also been facilitated by use of canals. Canals were first used by the early Assyrians, Egyptians, and Chinese for irrigation and later to transport their goods (USDA, 1955). The Grand Canal, built in China in the thirteenth century, served irrigation needs and provided an important artery of commerce for products of its millions of people. Other European countries, such as Holland, France, and England, later developed extensive systems of canals between natural waterways. In the Andes region of South America, rivers are also the arteries on which rubber, lumber, and other products of the interior can be carried to coastal areas.

During the several centuries of stability under the Roman Empire, vast and intricate systems of waterworks were constructed to provide millions of people with safe water (USDA, 1955). Disposal of sewage was well developed for the times, and, in general, the value of clean household water and of sanitation was well understood. However, when the empire collapsed, chaos reigned, and the excellent water system rapidly disintegrated. Constant warfare and political disturbances eroded the social underpinnings and values required to build and maintain an adequate water supply as a public service. As ignorance and poverty increased, sanitary precautions came to mean little and in time cleanliness was frowned upon as evidence of wicked thoughts and self-indulgence. Bathing, as formerly practiced for its therapeutic value, was abandoned. Citizens no longer took pride in having clean homes and streets. In fact, both became filthy. In addition, water obtained from wells became so fouled that it was unfit for human consumption (USDA, 1955).

Waterborne Diseases in Antiquity

Illness and death from waterborne diseases have plagued one country after another down to the present. Not only have poor people been struck down, but historical records have shown that many famous characters of history have fallen victims to waterborne diseases (USDA, 1955). Among them were King Louis VIII of France, Charles X of Sweden, Prince Albert of England, his son Edward VII, and his grandson George V. George Washington was also known to have suffered from dysentery, as did Abigail Adams, wife of Zachary Taylor, second President of the United States. Louis Pasteur's two daughters are said to have died of typhoid fever (USDA, 1955).

The popular opinion to maintain safe and clean water prevailed in the 19th century, especially in England and the United States. This was because dangers from contaminated supplies were generally known. In fact, the effects of contaminated waters, especially those used for drinking, now are considered to be a foremost priority in raising the basic standard of living in any developed country. The effects of contaminated water, especially groundwater, are just now, however, being fully realized by the general public. These effects will be discussed in more detail in later sections of this manuscript.

Qanats, the Great Waterworks

Qanats are some of the most extraordinary works of ancient man devised for collecting groundwater (Tolman, 1937). Qanats connect bottom shafts and are dug by laborers throughout a long period of time.

The qanats tunnels which supply Teheran are eight to sixteen miles long and reach 500 feet below the ground surface. One tunnel supplying a suburb of Teheran passes 200 feet below the city (Tolman, 1937; Wulff, 1968).

Another system of qanats supplies the city of Dizful (Tolman, 1937). The tunnels of this system extend under gravel bars of the river Abidiz, although a branch of qanats penetrate alluvial outwash gravels. The qanats of Dizful are also unusually productive and pass under the city to irrigate adjacent fertile land. The excavations extend two to six stories underground and are interconnected by subterranean passages. In this instance, qanats act as a source of water supply and sewage disposal for the dwellings.

ORIGINS OF QANAT BUILDINGS

The origin of qanat building is lost in antiquity, however, the tunnel of Negoub seems to have been built around 800 B.C. to supply the city of Ninereh (Butler, 1933). In 625 B.C. the Medes captured the city of Hamadan by destroying qanats supplying water to the city.

Water development by use of qanats was introduced by the naval captain, Scylax, in Egypt about 500 B.C., whom Butler (1933) designates as the "father of engineers." The qanats that were built in Egypt penetrate sandstone strata with water concentrated in faults. However, translations of old inscriptions and subsequent explorations revealed that they extend far to the east under a hundred miles of rolling desert to intercept seepage from the Nile (Tolman, 1937).

The art of qanat building is practiced by a guild whose members are known as "mukanni," under whose direction individuals perform manual labor. After exploratory shafts have been located a bed of water-bearing material and a series of construction shafts are dug. Shafts are sunk at varying distances at intervals of 100 yards. The qanats, in some instances, average sixty to the mile (Fraser, 1907). The grade and direction of the qanat from one shaft to another are determined by crude surveying instruments, consisting of a plumb line with a heavy weight hung from a tripod over the shaft (Tolman, 1937).

Iran, covering a large part of ancient Persia, developed this system for tapping underground water to a high level of sophistication (Wulff, 1968). The qanat water supply system in this area consists of underground channels conveying water from aquifers in highland areas to lower levels by gravity alone. The works built in Iran were built on a scale comparable to the aqueducts built by the Romans. The Iranian system is still in use after 3,000 years and in some instances has been expanded. In 1968, some 22,000 qanat units in Iran, comprising more than 170,000 miles of underground channels still existed (Wulff, 1968). The system supplied 75 percent of the water used and provided water for irrigation, and household consumption.

Wulff (1968) also described that during the seventh century B.C., the Assyrian King Sargon II reported he had observed an underground system for tapping water near Lake Urmia. His son, King Sennacherib, then applied the "secret" of using underground conduits in building irrigation systems around Nineveh. The son constructed a qanat system based on the Persian model to supply water for the city of Arbela. Egyptian inscriptions also disclose that Persians adopted the idea for Egypt after Darius I conquered it in 518 B.C. (Wulff, 1968). As noted above, Scylax, a captain in Darius' navy, built a qanat system that brought water to the oasis of Karg. This water was brought from the underground water table of the Nile River 100 miles away.

Qanats have been discovered throughout the cultural sphere of ancient Persia which includes Pakistan, Chinese oasis settlements of Turkistan, southern areas of the U.S.S.R., Iraq, Syria, Arabia, and Yemen (Wulff, 1968). During Roman and then Arabian domination of the area, they also spread westward to North Africa, Spain, and Sicily. In the Sahara region, oasis settlements were also irrigated by use of qanats.

Wulff (1968) provides an excellent description of the qanat system as it was observed in Iran during the 1960s. His account includes the use of burning oil lamps by the "mukanni." However, whatever the future of Iran's qanat system may be, it will be affected by the farmers who now drill wells and use diesel pumps.

Ancient River Civilizations and Use of Water

The concept of ancient river valley civilizations has been widely accepted (Coulborn, 1964; Smith, 1975) and five valley societies have been extensively known. These include: Egypt in the Nile Valley; Mesopotamia in the Tigris-Euphrates Valley; Indian civilizations in the Indus Valley; China in the Yellow River Valley; and the Andean civilization in river valleys of coastal Peru. To what degree these groups practiced and perfected irrigation technology is difficult to discern. However, some small-scale irrigation descriptions, perhaps on the tributaries of a big river system, are plausible (Smith, 1975).

Other Related Information

In Egypt, irrigation was dictated and conditioned by the Nile (Smith, 1975). Of overriding importance was the fact that once a year the river flooded, regularly and predictably. Flooding began in August and reached a peak at the end of the month. It then receded two months later. Extending the area of fertility by cutting irrigation channels to increase domestic supply during the flood stage seemed logical.

In the forested regions of northwestern, northern and central Europe, conduits which carried water were made of wood (Clark, 1944). Channels were cut from solid split timbers and built up from boards. In some instances whole trunks were also hollowed into the shape of tubular pipes. Timber pipes, often with iron hoops at their joints, were then used for distributing water to most of the cities of northwestern Europe. This has occurred until recent times.

Rainwater as a Source of Water

Rainwater has been collected from rooftops and from flights of steps (Clark, 1944). In areas short of water, rainwater collected via this fashion formed a significant source of supply. Rainwater, especially in areas where acid rain does not occur, was valuable for industrial purposes. This even occurred when other domestic sources of water were available. The Romans also knew this, for they built wood-lined cisterns in their posts. They adequately stocked themselves with well water, usually for drinking, however, storage of this type could have also been used for industrial purposes (Clark, 1944). Ancient civilizations also esteemed rainwater for certain purposes, possibly owing to its softness and freedom from salts commonly dissolved out of rock formations through which well water passed. Rainwater would be useful for dyeing fabrics, washing linen, and preparing flax.

Summary of a Historical Perspective on Water

An evaluation of the history of water, its sources, and uses reveal an important characteristic that is fundamentally important throughout history. Its use has always been necessary for technical, organizational, managerial, and social requirements. This characteristic becomes even more complex when toxic substances and contamination of certain water resources because of these activities are considered.

Water for drinking has always been a primary need of humans since the formation of the first urbanized communities. In fact, in many portions of the world, successful agriculture depends on irrigation, a technology whose ramifications extend beyond raising of crops. The application of water for electrical and power generation has also been a powerful determinant in the evolution of technologies in both the eastern and western hemispheres. Ultimately this technology has influenced the spread and development of industrialization. Hydro-technology, however, is only one facet of the total relationship that exists between water and man. The development of methods to capture, control, distribute, and use water have been the predominant issues.

The twin problems of providing adequate water quantities and ensuring its quality pose a significant challenge to water resource managers and workers. Water quantity and quality raise political, economic, environmental and social issues. These issues at present are of such considerable magnitude that they will undoubtedly become more important in the years to come. Environmental contaminants (e.g., toxic chemicals) and water relationships will continue to be big issues. This change from specifically oriented engineering problems towards those of water resources and their proper management represents the basic distinction between past and present hydro-technology. Today the problem is one in which adequate quality of water must be maintained. Sufficient quantities of water for domestic consumption must also be supplied.

Raw water can contain substantial quantities of natural, human, agricultural and industrial contaminants. These in turn can reduce it to a level unsafe to drink. Treatment to remove impurities is also complex and expensive. Many water-treatment plants are by no means equipped to eliminate obnoxious contaminants which have appeared over the last few years. Among some of the contaminants that *must* be removed are: phenols, high levels of "hard" detergents, radioactivity, a variety of pesticides, fungicides, and insecticides. All of the previously mentioned can and do find their way into groundwater sources.

Technologically, therefore, man must meet the demands placed on him in reference to removing contaminants from water. If the human species fails to take steps now that will ensure an ample supply of fresh groundwater tomorrow, the failure will be one of management. This view of course must not be allowed to take hold. We must continue, as stated by Clark (1944):

> So, from the Stone Age to the twentieth century has water reflected the image of society; first the food-gathering tribes camping round their springs; then the lowly peasants with their village tites; next the town-dweller and his aqueducts and wells with their diverse appliances and the water-driven mills and lathes; and finally the men of the future, breaking loose from the city bounds, banishing smoke and fumes and winning leisure for themselves and the control of energy for society at large.

Protection of a water supply system and its management must also include other things (Rail, 1985a):

> Effective protection of groundwater against undue contamination or pollution will require a well-integrated monitoring plan at the local, state, and federal levels, and the use of a groundwater monitoring protocol . . . can have the benefit of providing a safe and ample supply of water for a community. Its use can also assist in centralizing water quality and quantity data that is presently scattered among many water-related agencies. The collected and summarized data can then form the nucleus of a water data bank from which information can be shared among the participating agencies within the groundwater monitoring network. The information would be a significant asset for water management purposes.

TWO

The Natural Quality of Water

The natural quality of groundwater is affected by many things. Among the most important relating to chemical and microbiological changes are geological characteristics and climatic conditions (e.g., temperature, rainfall; Pye et al., 1983; Hem, 1959, 1970; Matthess, 1982).

Geological Effects and Natural Quality

The ultimate source of most dissolved ions in water is the mineral assemblage which occurs in rocks near land surfaces (Hem, 1970). The solids which dissolve into groundwater from this point, form part of the geochemical cycle. This starts from the stock of material in igneous rocks, proceeds via weathering, forms sedimentary rocks, moves through terrestrial waters, and eventually to the sea (Matthess, 1982). An important factor is geochemical mobility of elements in a geochemical environment (e.g., in rocks, the hydrosphere, the atmosphere). Their solubility (i.e., the amount of a substance that will dissolve in another substance) is also an important factor.

Rock composition, purity, and crystal size of its minerals, rock texture and porosity, regional structure, degree of fissuring, length of previous exposure time, also influence the composition of water passing over and through rocks (Hem, 1970). The natural chemical content of groundwater is consequently influenced by type, depth of soils, and subsurface geologic formations through which groundwater passes (Pye et al., 1983). Leaching of soil and rocks are also influenced by the acidic, neutral, or alkaline pH conditions of precipitation, and metabolic products of organisms which occur.

Besides the fact that some of the elements are soluble, there is a tendency for them to become incorporated into other new minerals, be adsorbed (i.e., be taken up and held) on active surfaces, or to be co-precipitated with others. These tendencies must be considered when evaluating the mobility of elements (Matthess, 1982). An increase in temperature will also raise the solubility of inorganic solutes (i.e., matter other than plants or animals) and the rate of dissolution of rock minerals (Hem, 1970). Since most thermal groundwaters (e.g., hot springs) occur in areas where temperature gradients with depth are relatively steep, they may be high in dissolved solids and contain unusual amounts of metallic ions.

Water entrapped in igneous rock that is released on fusion of the rock minerals can be of a substantial quantity (Clarke and Washington, 1924). It is estimated that over one percent of the weight of an average igneous rock can be entrapped water. A substantial amount of water could be held in this entrapped form within igneous rocks located on the crust of the earth, although there is some doubt as to the validity of this figure (Goldschmidtt, 1954). White and Waring (1963), however, have also shown water to be very predominant in volcanic gases.

Within an aquifer, water contacts geologic material and this reaction usually increases its content of total dissolved solids. Naturally occurring radionuclides, if they occur in geological formations, may render the water radioactive. The hardness factor in water reflects calcium and magnesium content and at certain levels renders water unfit for human consumption, whether it be domestic or industrial use. In fact, water from some aquifers may be unusable because of high salinity or the presence of high levels of naturally occurring toxic substances such as arsenic and radionuclides. Groundwater in its natural state may be unfit for certain uses or be of a high quality that also can be used for drinking water without any treatment required.

Inorganics

Substances occurring in water can be classified into different degrees of mobility (Matthess, 1982). These groupings include:

- extremely mobile substances, such as chlorides and sulfates

- somewhat mobile substances, such as calcium, magnesium, and sodium
- generally mobile substances, such as silicon, phosphorus, and potassium
- slightly mobile and inert substances such as iron, chromium, aluminum, and lead

Common elements in igneous rocks (e.g., silicon, aluminum, iron) have little mobility within the hydrosphere (Matthess, 1982). However, chlorine, which is relatively scarce in the earth's crust, is very mobile and widespread in the hydrosphere. Calcium is mobile and sodium has considerably more mobility than potassium, although both elements occur in approximately equal amounts within primary igneous material (Matthess, 1982).

It appears that the mobility of an element in the hydrosphere is determined by its solubility and the various compounds it forms, including the tendency of its ions to deposit on rocks.

Geochemical Investigations

A general understanding of the composition of rock materials is necessary to understand chemical compositions of natural water (Hem, 1970). Summaries of many analyses have been made by Clarke and Washington (1924), Clarke (1924), Fleischer (1953, 1954), Turekian and Wedepohl (1961), Taylor (1964), and Parker (1967). Oxygen is the most abundant of the elements in crustal rocks (Goldschmidtt, 1954). According to Clarke and Washington (1924), 95 percent of the crust of the earth, at least to a depth of ten miles, is igneous rock. However, Hem (1970), maintains that when considering natural water and rock composition, this predominance of igneous rock is basically not important. This is because most recoverable groundwaters occur at depths of less than one mile below the land surface. Also, in the portion of the earth's crust near surface areas, sedimentary rocks are more prevalent than igneous rocks. Hem (1970) argues that igneous rocks make poor aquifers and, consequently, transmit little water.

In headwater areas of many streams, igneous rocks occur at the surface and contribute solutes to surface runoff. This is done by direct contact and through leaching of decomposed minerals in overlying soil. According to Hem (1970), these areas are not a predominant part of the earth's surface. Thus, sedimentary rocks and soil areas assume major importance as the immediate source of soluble matter that can be taken up by circulating ground and surface waters. However, reactions between water and minerals of igneous rocks are still important. These reactions must be considered in any groundwater quality evaluations.

Natural Water Chemistry

Principles and processes controlling the chemical composition of natural water have been published (Garrels and Christ, 1964; Krauskopf, 1967; Hem, 1959, 1970; Matthess, 1982). These previously listed publications are concerned with the behavior of minerals and metallic elements in aqueous environments. They also apply directly to natural water chemistry, however, a general summary must be presented prior to any discussion of the subject. The following summary material is presented in a brief format, but the references cited allow one to pursue selected subjects further.

Summary of Theoretics on Natural Water

The theoretical basis of the study of natural water chemistry consists of the following fundamental factors (Hem, 1970):

- fundamental alterations of atoms and nuclei
- various kinds of chemical reactions involving water, solutes, solids, or gases
- methodology used to study these reactions systematically, specifically, as water moves through the hydrologic cycle

Other studies concerning the understanding of the chemistry of natural water stem from the large amounts of work expedited when initially prospecting for mineral deposits (Hem, 1970). Additional pioneering work on natural water composition and mineral prospecting was also done in the U.S.S.R. (Fersman et al., 1939; from Hem, 1970). Hem (1970) cities numerous studies both in the United States and abroad which directly relate to natural water chemistry (Valyashko, 1958, 1967; Chebotarev, 1955). Hem (1970), also lists many aspects of natural water chemistry and includes sections on: types of reactions in water chemistry, chemical equilibrium, temperature and pressure effects, application of chemical thermodynamics, electrochemical equilibrium, chemical reaction rates, solubility concepts, and adsorption and ion-exchange calculations. Information on geologic effects, climate, biochemical factors, and ecology applied to natural water, is also included.

Aquifer Materials and Groundwater Quality

According to Matthess (1982) groundwater changes during its passage through an aquifer because dissolution, hydrolysis, precipitation, adsorption, ion exchange, oxidation, reduction, and microbial metabolism are always occurring to some degree. Consequently, variations occur among multiple aquifers within similar rock masses and this can lead to classification of groundwater by various types.

Igneous Rocks and Natural Water

In considering igneous rocks in relationship to water composition, texture and structure of the rocks are significant as they relate to the surface area of solid rocks that may be exposed to attack. Groundwater may also be recovered in large amounts from some of the extrusive igneous rocks where there are shrinkage cracks and other joints, interflow zones, or other openings through which water may move (Hem, 1970). Most igneous rocks, however, are impermeable. Surface water originating in areas where igneous rocks are exposed is low in dissolved solids because generally weathering action on igneous rocks is slow. The concentrations of dissolved solids eventually reached are likely to be a function of contact time and area of solid surface exposed to water. In addition, where plant growth is occurring, an enhanced supply of carbon dioxide occurs and hydrogen ions become available to circulating water.

Sedimentary Rocks and Natural Water

Sedimentary rocks classified as resistates have a considerable range of chemical composition (Hem, 1970). Consolidated resistate sediments, such as sandstones, contain cementing material deposited on the grain surfaces and within openings and grains. This cementing material is deposited from water that has passed through the rock at some past time and can be redissolved. Common cementing materials include calcium carbonate, silica, ferric hydroxide, or ferrous carbonate, silicate, ferric hydroxide, or ferrous carbonate. However, other materials such as clay can occur in composition.

Resistate rocks can have their chemical composition affected in many ways, consequently, water from these rocks displays a wide range of chemical quality. This chemical quality, however, can only be determined by actual sampling and analysis of a specific aquifer, although it can be predicted in advance.

Sedimentary rocks including shale and other fine-grained rocks (hydrolyzates) are composed in large part, of clay minerals and other particulate matter formed by chemical reactions between water and silicates (Hem, 1970). Although shale and similar rocks are porous they do not transmit water because openings are small and poorly interconnected. The water, consequently, that can be obtained from a hydrolyzate rock contains high concentrations of dissolved solids. Hem (1970) presents an excellent discussion, and some examples of hydrolyzates and natural water complexities.

Sedimentary rocks termed precipitates include carbonates, e.g., limestone, which is calcium carbonate, and dolomite, which is principally calcium and magnesium (Hem, 1970).

Metamorphic Rocks and Natural Water

Rocks of any kind can be metamorphosed and the process consists of the alteration of rock by heat and pressure to change the physical properties and mineral compositions (Hem, 1970). Many associations of metamorphic rocks and their elements occur because of the wide variety of original rocks.

For example, the dense structure of slate (i.e., a metamorphosed form of shale) and quartzite (i.e., a metamorphosed form of quartzose sandstone) restrict water movement toward fractured zones and prevent groundwater from contacting large surface areas of minerals in shale or sandstone areas (Hem, 1970). The opportunity for groundwater then, to pick up solutes from metamorphic rocks is small.

Metamorphism generally involves aqueous solutions and the resultant alteration products include hydrolyzate minerals (Hem, 1970). The release of water from various types of rocks during metamorphic processes can also be expected. White (1957, 1960) examined natural waters for possible features of their composition that could be attributed to metamorphism of rocks. Among properties that were cited as indicators of metamorphic influence on water composition were high concentrations of sodium, bicarbonate, boron, and low concentrations of chloride.

Dissolved Constituents in Groundwater

Total Dissolved Solids

The U.S. Public Health Service (U.S. Public Health Service, 1962) standard for total dissolved solids (TDS) in drinking water is 500 ppm, although the 1974 Safe Drinking Water Act (SDWA, 1974) considers waters containing up to 10,000 ppm, potential sources of drinking water (U.S. EPA, 1977). Total dissolved solids comprise dissociated and undissociated substances, but not suspended material, colloids, or dissolved gases (Matthess, 1982). The four types of natural groundwater that exceed 10,000 ppm TDS are connate water, intruded seawater, magmatic and geothermal water, and water affected by salt leaching (including the products of evapotranspiration) (Pye et al., 1983). The TDS value is determined by evaporation of a known sample of water to dryness (i.e., 1–2 hours drying time at 180°C). The residue on evaporation, however, is not the same as the initial solution content, with hydrogen carbonate ions being precipitated as carbonates, and sulfate ions as gypsum (Matthess, 1982). Small amounts of magnesium, chloride, nitrate, and boron, as well as organic substances, usually escape. If the loss of carbon dioxide on precipitation of carbonates is taken into account by adding half the hydrogen carbonate ion content to the residue on evaporation, the result is then reported as "total."

In groundwater areas, water reacts with geologic formations increasing the content of dissolved solids (Pye et al., 1983). Thus, the natural chemical quality of groundwater depends upon age and geological formations encountered throughout its flow history. Carbonate geological formations would increase magnesium and calcium content of the groundwater and the concentration of bicarbonate ions would also increase due to the dissolution of calcite and dolomite. Alumina silicate minerals would, under similar circumstances, increase concentrations of sodium, potassium, magnesium, calcium, and silicon hydroxide (Pye et al., 1983) in groundwater.

The basic chemical constituency of water at a particular point in time in consolidated deposits (e.g., sedimentary, igneous, and metamorphic sources) is based on characteristics of the different assemblages (Pye et al., 1983). For example, sulfate bearing minerals (e.g., gypsum and anyhydrite) are characterized by having high solubilities. Consequently, in older groundwater that has encountered sulfur bearing minerals, sulfate anions dominate bicarbonate anions. In deep, considerably old groundwater the chloride ions dominate both bicarbonate and sulfate ones due to the presence of soluble minerals (e.g., halite, NaCl, sylvite, and KCl) (Pye et al., 1983).

NATURE OF DISSOLVED IONS IN GROUNDWATER

Solutions are described as uniform mixtures in which dissolved particles are surrounded by solvent molecules (Hem, 1959, 1970). These molecules have no direct contact with other like particles, except when movements occur through the solution. As individual ions or pairs of ions, and/or complexes made up of several ions, they can be considered to be in the dissolved state. However, once particles form aggregates and settle out by gravitational action, they are no longer considered to be in the dissolved state.

Natural groundwater in some areas carries dissolved and suspended particles. The amounts of suspended particles present in natural groundwater, however, is small (Hem, 1959, 1970). The total concentration of dissolved material in groundwater is determined from the weight of dry residue remaining after evaporation of the volatile portion of the water sample. Total dissolved solids or dissolved solids are terms used synonymously to describe this value.

Organic matter in water samples may be partly volatile but cannot be removed unless the residue is ignited. Inorganics such as nitrate and boron are partly volatile, and water that has a low pH generally loses a considerable amount of its anion content when evaporated to dryness (Hem, 1970). Waters high in sulfate (e.g., crystals of gypsum), on the other extreme, contain water of crystallization that is difficult to remove at a 180°C drying temperature.

TOTAL DISSOLVED SOLIDS AND SIGNIFICANCE

Total dissolved solids content amounts to less than 10 mg/l in rain and snow, less than 25 mg/l in water from humid regions, and more than 300,000 mg/l in brines (Matthess, 1982; Hem, 1959, 1970; Davis and DeWiest, 1967). Waters of high dissolved solids have been assigned terms by the U.S. Geological Survey (Hem, 1970; Robinove et al., 1958):

	Dissolved Solids (mg/l)
Slightly saline	1,000–3,000
Moderately saline	3,000–10,000
Very saline	10,000–35,000
Briny	>35,000

In developing standardized procedures for computing the total dissolved constituents the assumption was made that results will be interpreted as a substitute for the determined residue on evaporation as previously described. The value obtained is meant to correspond to conditions which exist in the dry residue state. This assumption is not always correct, but has been a good approximation of dissolved solids (Hem, 1970).

Inorganic Ions

SODIUM

Sodium does not occur free in nature, but compounds of this element constitute 2.83 percent of the earth's crust (McKee and Wolf, 1963). Sodium salts are extremely soluble in water and when the element is leached from soil or discharged to streams by industrial waste processes, it remains in solution. Sodium is a cation of many salts used in industry and therefore is common in process wastes.

Sodium, if present in high amounts in drinking water may be harmful to individuals with cardiac, renal, and circulatory diseases. As little as 200 mg of the element in drinking water may be injurious to health (McKee and Wolf, 1963; Muhlberger, 1950; Laubusch and McCammon, 1955; AWWA, 1950). Excess concentrations of sodium salts in drinking water can also be deleterious to animals. Well waters containing 7000 mg/l of sodium have been known to be toxic to chicks (Tyler, 1949). A threshold upper limit of 2000 mg/l of sodium that would cause problems for livestock, was also suggested by Stander (1961; from McKee and Wolf, 1963).

Sodium, a major constituent of igneous rocks, occurs as plagioclase, i.e., the soda-lime feldspar with the sodium-rich member albite (Matthess, 1982). Of less importance are sodium-containing silicates such as nepheline, sodalite, jadeite, arfredsonite, glaucophane, aegirine, and members of the zeolite group such as natrolite. Sodium is liberated during the weathering of these silicates, and generally, the distribution of this element and the rate of its liberation is reflected in the chemical quality of spring water.

Sodium salts and their high solubility, including their degree of sorptive bonding onto clay minerals, considerably enrich seawater. Concentrations of 10,800 mg/kg can occur in seawater and in evaporative deposits where it chiefly shows up in rock salt, NaCl (Matthess, 1982). In sandstones, the element averages 3870 mg/kg concentration, either as a constituent of unweathered mineral grains, or as a constituent of the cement. Sodium in the presence of prolonged or quantitatively lower groundwater movement can be the cause of higher salt concentrations, which are generally leached out (Matthess, 1982). Hydrolysates (argillites) are also sodium rich (i.e., 4850 mg/kg), partly as a result of the lower permeability of this fine-grained material. The sorptive bonding of sodium to clay minerals would also contribute to this.

Sodium also occurs as a cyclic salt or as terrestrial dust in small amounts (0.2 mg/l; Matthess, 1982). Sodium as a bioelement can also occur in chemicals used by man. Consequently, it enters into groundwater because of human pollution activities.

Sodium is usually present in freshwater as Na^+ ions within a concentrated solution (Hem, 1959, 1970). Complex ions and sodium ion pairs, such as sodium carbonate, sodium bicarbonate, and sodium sulfate can also occur in these solutions. The solubility of sodium salts is high and a pure solution of sodium hydrogen carbonate contains 15,000 mg Na^+/l at room temperature. Occasionally the solubility limit of sodium salts can be exceeded, particularly in inland drainage basins. Here it can lead to precipitation of sodium salts, e.g., sodium carbonate, sodium sulfate, sodium nitrate (Langbein, 1961; Toth, 1966; Matthess, 1982). The highest sodium concentrations, however, occur in association with Cl^- ions (Hem, 1959, 1970). The saturation concentration in relation to NaCl occurs at approximately 150,000 mg Na^+/l. High concentrations of this order are usually not reached in natural waters, even in rock salt-bearing strata. The highly concentrated solutions originating on contact surfaces are removed continuously by the flow of water.

The highly sodium-rich groundwater which occurs in rock salt-bearing strata (i.e., Salado Formation and Rustler Formation in Eddy County, New Mexico, at 97 m depth) contained 121,000 mg Na^+/l (Hem, 1970). This gave a total dissolved solids content of 329,000 mg/l.

Sodium is the dominant cation in mineralized groundwaters excluding salt-bearing sediments and evaporites. The sodium content of groundwater in humid climates is of the order of 1–20 mg/l (Matthess, 1982).

Irrigation water with high sodium levels can bring about a displacement of exchangeable cations (i.e., Ca^{+2} and Mg^{+2} from clay minerals in soil) followed by replacement of the cations by sodium (Matthess, 1982). Sodium saturated soils can also peptize and lose permeability functions, so that their suitability for cultivation decreases. However, the process is reversible and can be counteracted by addition of gypsum.

The ratio of Na^+ ions to total cation content (i.e., as stated in meq % in conjunction with total dissolved solids or electrical conductance) can be used for assessment of the suitability of water for irrigation (Todd, 1960; Matthess, 1982).

The ability of water to expel calcium and magnesium by sodium can be estimated via use of the Sodium Absorption

Ratio (SAR) (Richards, 1954; Hem, 1959, 1970). High SAR values indicate the risk of displacement of alkaline earths.

There also seems to be a risk from water with a high hydrogen carbonate content and low calcium content (Matthess, 1982). Consequently, water with more than 2.5 meq/l residual sodium carbonate should not be used; 1.25–2.5 meq/l should not be used if possible; and water with less than 1.25 meq/l may be regarded as safe. Some Figures and Tables depicting sodium levels in the Albuquerque, New Mexico area (Rail, 1986) are shown in Appendix A of this manuscript. Other inorganic chemicals and water quality summarization of data are also in Appendix A.

POTASSIUM

The average potassium content of igneous rock is 25,700 mg/kg (Matthess, 1982). Potassium is also abundant in sedimentary rocks (Hem, 1970). Potassium exhibits a strong tendency to be reincorporated into solid weathering products, specifically, clay minerals. Potassium occurs in potassium feldspars (e.g., orthoclase and microcline), the micas (e.g., muscovite and biotite), and in other minerals (e.g., leucite, nepheline) (Matthess, 1982). Potassium ions are released by weathering, and according to Matthess (1982), can become fixed again through sorption on clay minerals.

The Na^+/K ratio in igneous rocks is 1.09 and 27.84 in seawater (Matthess, 1982; Rankama and Sahama, 1960). High potassium levels of 13,200 mg/kg in sandstone are caused by absorbed potassium in the cementing material and by unweathered portions of potassium feldspar and micas (Hem, 1970). Evaporite rocks include beds of potassium salts and also constitute a high source for potassium concentration in brines.

Potassium (K) is an essential plant nutrient and also shows characteristics of having its own biological cycle (Hem, 1970; Matthess, 1982). Incorporation of the element by vegetation draws it from the hydrologic cycle. However, the amounts involved can hardly be responsible for the significantly lower potassium content in groundwater relative to sodium. This is because the materials incorporated into plants are once more returned to the soil by plant decay and are there subjected to possible leaching (Matthess, 1982).

Potassium is also used as an agricultural fertilizer and when used excessively via this method leads to higher concentrations in groundwater (Harth, 1965; from Matthess, 1982). The solubility of the potassium salts in water is high and at 20°C water saturated with KCl can contain approximately 133,000 mg K^+/l of the element (Matthess, 1982; Davis and DeWiest, 1967). High potassium values generally are also associated with comparable levels of human pollution (Matthess, 1982).

Potassium seldom occurs in greater or almost equal concentrations when compared with sodium because of its lower geochemical mobility in freshwater (Matthess, 1982). In general, the potassium concentration rises with increasing mineral matter content more slowly. Consequently, groundwater with 2000 mg/l total dissolved solids often contains less than 20 mg K^+/l.

Potassium is also an active metal and reacts vigorously with oxygen and water (McKee and Wolf, 1972). Potassium salts are also substituted for sodium salts in many industrial operations. Sodium salts, however, are generally and more frequently used because they are less expensive. Potassium salts are also used for fertilizers and some varieties of glass (McCutcheon et al., 1936).

Potassium is an essential nutritional element, however, if present in excessive quantities it is a cathartic (i.e., at doses of 1 to 2 g). The element at levels of 1000 to 2000 mg/l is regarded as the extreme light that should be allowed in drinking water (McKee and Wolf, 1972).

CALCIUM

Calcium does not occur naturally because it oxidizes readily when in contact with air and reacts with water to release hydrogen gas (McKee and Wolf, 1972). Calcium salts, however, are common in water. These salts may result from leaching of soil and other natural sources or they may occur in sewage and industrial wastes.

Calcium is also the principal cation in freshwater (Hem, 1970). The element is widely distributed in common minerals of rocks and solids, and it is an essential constituent of igneous rocks (i.e., especially chain silicates, pyroxene and amphibole, and feldspars; Hem, 1970). The average content of the element in igneous rocks is 36,200 mg/kg. It is held in plagioclase feldspar, a solid solution series with end members anorthite and albite. There is also an appreciable calcium content in associated amphibole and pyroxene mineral groups and in garnet groups (Matthess, 1982).

Calcium on weathering and deposition is incorporated in resistates (22,400 mg/kg), or in hydrolysates (22,500 mg/kg) (Matthess, 1982). The chief concentration of the element occurs in precipitates, specifically, carbonate rocks (272,000 mg/kg). Calcium also forms deposits of carbonates, calcite, aragonite, dolomite, and the sulfates anhydride and gypsum. The element also is present in the form of adsorbed ions on negatively charged mineral surfaces in soils and in rocks (Hem, 1970).

Calcium ions are relatively large and can be hydrated (Matthess, 1982). They form complexes with other inorganic ions and affect significantly the calcium concentration in natural waters in specific instances. Strong alkaline waters also contain hydroxide and carbonate ion pairs with calcium (Hem, 1970). Excessive calcium levels in drinking water have under certain conditions also been implicated as factors relating to the formation of kidney or bladder stones in the human body (McKee and Wolf, 1972; Kovalev, 1954).

MAGNESIUM

The behavior of magnesium is similar to calcium, and the two elements form the property classified as hardness (Hem, 1970). The geochemical behavior of magnesium, however, is different from calcium. Magnesium ions are smaller than sodium or calcium ions and they exhibit a stronger attraction for water molecules.

The average magnesium content of igneous rocks is 17,600 mg/kg (Matthess, 1982). Magnesium-containing minerals include: olivine, forsterite, garnets, cordierite, pyroxene, amphibolite, mica groups, chrysotile, sepiolite, talc, serpentine, chlorite groups, and magnesium-bearing clay minerals (Matthess, 1982). Significant amounts of magnesium are also contained in precipitates (45,300 mk/kg), in carbonate dolomite, magnesite, and mixtures of these which occur in limestone. Magnesium is also a constituent of salt deposits of marine origin (e.g., bischofite, kieserite, hexahydrite, epsomite, carnollitte), and the salts are also present in terrestrial evaporites and in guano (Toth, 1966; Rankama and Sahama, 1960). The element comes from the ocean, from dust of magnesium-containing rocks, or from industrial emissions (Matthess, 1982).

The solubility of magnesium carbonate in water and carbon dioxide is greater than calcium carbonate (i.e., so that under normal groundwater conditions magnesium carbonate is not precipitated) (Hem, 1970). Magnesium levels in freshwater are generally less than calcium, probably because of the lower geochemical abundance of magnesium (Matthess, 1982). Freshwater usually has magnesium levels below 40 mg, although higher concentrations will be detected in magnesium rocks that have high salt content, e.g., 51,500 mg Mg/l in a well in Eddy County, New Mexico with 436,000 mg/l total dissolved solids (Hem, 1970).

Magnesium is also distributed in many ores and minerals (Anonymous, 1960; McKee and Wolf, 1972). However, because it is active chemically, it does not occur in a natural elemental state (i.e., its salts are very soluble; Anonymous, 1950, 1950b). Magnesium ions can be of importance in water contamination evaluations if they occur in high concentrations in natural waters. Magnesium and calcium form the bulk of the hardness reaction.

Magnesium is also an essential mineral element for humans, with daily requirements being 0.7 g (Shohl, 1939). Magnesium is non-toxic to humans and does not constitute a public health hazard because, before toxic levels occur in water, the taste cannot be tolerated (Negus, 1938). Magnesium, however, at high levels has laxative effects, particularly upon those who have never ingested it excessively (McKee and Wolf, 1972).

The 1946 USPHS Drinking Water Standards recommended a limit of 125 mg/l for magnesium; however, none is given in the 1962 standards (USPHS, 1962). The 1958 WHO International Standards (WHO, 1958) listed a "permissible limit" of 50 mg/l and an "excessive limit" of 150 mg/l, but no maximum allowable concentration.

Magnesium and calcium in the animal body are antagonistic and calcium appears to alleviate symptoms of magnesium excess. Diets high in magnesium and low in calcium cause ricketts (Shohl, 1939; McKee and Wolf, 1972).

Magnesium salts also act as cathartics and diuretics among animals and human beings. High concentrations of magnesium in drinking water cause scouring diseases in livestock (McKee and Wolf, 1972). Ingestion of mixtures of sodium salts and magnesium with nitrate ions has also caused poisoning among ducks (Heller, 1933). Concentration levels of magnesium and calcium in water are a factor controlling the distribution of certain crustacean fishfood organisms, i.e., copepods in streams (Hedgepeth, 1944).

HARDNESS

Hardness in water is caused by metallic ions dissolved in water (U.S. EPA, 1976). In fresh waters these ions include calcium and magnesium ions, although iron, strontium, and manganese also contribute to hardness if appreciable concentrations occur. Hardness is also commonly reported as an equivalent concentration of calcium carbonate.

Frequently, the influence of water hardness, certain minerals or heavy metals contained in drinking water on morbidity and mortality is placed in doubt (Sonneborn et al., 1983). Applying the same methodological criteria, all large scale studies carried out in Japan, the U.S.A., England, Canada, Sweden, and Finland, have shown that there are statistical interrelationships between the degree of regional mortality due to cardiovascular diseases and certain ingredients of the drinking water of the same regions. However, these interrelationships appear to be fortuitous. In some studies, they refer to water hardness, in others to calcium or magnesium, and to a varying spectrum of other ions, heavy metals, and trace elements.

The concept of hardness, however, initially came from the water supply practice field and was measured by soap requirements needed for adequate lather formation. Hardness was also used as an indicator of the rate of scale formation in hot water heaters and low pressure boilers. Sawyer (1960) classified hardness as follows:

Concentration (mg/l Calcium Carbonate)	Description
0–75	soft
75–150	moderately hard
150–300	hard
300+	very hard

A natural source of hardness includes limestones which are dissolved by percolating rainwater rendered acidic by dissolved carbon dioxide (U.S. EPA, 1976). Industrially related sources of hardness include inorganic chemical industries and discharges which occur from operating and abandoning mines.

Durfor and Becker (1964) presented the following criteria for hardness:

Range (mg/l Calcium Carbonate)	Description
0–60	soft
61–120	moderately hard
121–180	hard
>180	very hard

The United States PHS Standards (USPHS, 1962) does not specify values for hardness. The American Water Works Association, however, indicates that "ideal" quality water should not contain more than 80 mg/l of hardness (Bean, 1962).

Hardness in water when used for ordinary domestic purposes is not objectionable until a level of 100 mg/l is reached (Hem, 1970). Hardness in many instances exceeds this level, especially where waters have contacted limestone or gypsum (i.e., 200–300 mg/l is common). Because hardness is a property not attributable to a single constituent, some convention has to be used for expressing concentrations in quantitive terms (Hem, 1970). Hardness therefore is reported as equivalent concentrations of calcium carbonate (Hem, 1970).

Hardness is computed by multiplying the sum of milliequivalents per liter of calcium and magnesium by 50. The hardness value resulting is entitled "hardness as calcium carbonate." The same quantity can also be referred to as "calcium + magnesium hardness" or "total hardness." If incomplete water analyses are available, total hardness may serve as a measure of "total dissolved solids" for freshwater (Matthess, 1982). The residue after evaporation of the water increases linearly with hardness.

The National Academy of Science (NAS, 1974) panel recommended against the use of the term *hardness* and suggested that the specific ions involved be used. For most existing data, it was difficult to ascertain whether toxicity of various ions was reduced because of the formation of metallic hydroxides and carbonates caused by associated increases in alkalinity, or because of an antagonistic effect of principal cations contributing to hardness (e.g., calcium, magnesium, iron, manganese, copper, barium, zinc) (U.S. EPA, 1976; Matthess, 1982). Stiff (1971) in reference to this, presented a hypothesis which explained if cupric ions were the toxic form of copper, then observed that differences in toxicity of copper between hard and soft waters could be explained by the difference in alkalinity rather than hardness. On the opposite extreme, Doudoroff and Katz (1953) presented data showing that increasing calcium reduced the toxicity of some heavy metals. The observed reduction of toxicity of metals in waters containing carbonate hardness as measured by a 40-hour LC-50 for rainbow trout was estimated to be four times for copper and zinc. This occurred when hardness values were increased from 10 to 100 mg/l as calcium carbonate (NAS, 1974).

Hem (1970) and Muss (1962) summarized statistical correlations between hardness, the properties of drinking water supplies and death rates from cardiovascular diseases. Muss (1962) observed that lower death rates from heart and circulatory diseases occurred in states where public water supplies were higher in hardness.

IRON

By weight, iron is the fourth most abundant of the elements that make up the earth's crust (U.S. EPA, 1976). It is common in many rocks and is a component of many soils, especially clay soils. Iron in groundwater is present in varying quantities, which depend upon the geology of the area and other chemical components of the water source.

Iron is also an essential trace element required by plants and animals (U.S. EPA, 1976), and can be a limiting factor for the growth of algae and other plants, and can limit growth of organisms in marl lakes. In igneous rocks the principal minerals that contain iron as an essential component, include the pyroxenes, amphiboles, biotite, magnetite, and the nesosilicates (Hem, 1970). However, most commonly, iron in igneous rocks occurs in the ferrous form, mixed with ferric iron as magnetite (Hem, 1970).

Iron also occurs in garnets such as andradite and almandite (Matthess, 1982). Occurrence of the element in groundwater is influenced by microorganisms which assist in its oxidation to ferric ion (under aerobic conditions) or its reduction to divalent iron (under anaerobic conditions). Also, according to Matthess (1982) an increase in divalent iron content can be used as a criterion for the presence of groundwater contamination by organic substances.

Water of low pH (i.e., which may result from discharge of industrial wastes, drainage from mines, or from thermal springs) can also carry high concentrations of ferric and ferrous iron (Hem, 1970). Zelenou (1958; from Hem, 1970) also noted that large amounts of iron may be discharged by hot springs in volcanic areas.

The ferrous, or bivalent (Fe^{++}), and the ferric, or trivalent (Fe^{+++}) ions, are the primary forms of concern in the aquatic environment, although other forms may occur in organic and inorganic wastewater streams (U.S. EPA, 1976). The ferrous form can persist in waters void of dissolved oxygen and originates usually from groundwaters or mines whenever they are pumped or drained.

Significant iron contamination sources include industrial wastes, mine drainage waters, and iron-bearing groundwaters (U.S. EPA, 1976). Iron in water from mine drainage is precipitated as a hydroxide in the presence of dissolved oxygen. These yellowish deposits, called "yellow boy," occur in streams draining coal mining regions in Appalachia. Occasionally ferric oxide is precipitated, which reddens water.

High iron levels are an objectionable constituent in water supplies used for domestic or industrial purposes. Iron in high concentrations can affect the taste of beverages and can stain laundered clothes and plumbing fixtures. Iron in spring water can also be tasted at levels of 1.8 mg/l and in distilled water at 3.4 mg/l (Cohen, 1960).

The daily requirement for iron in human individuals is 1–2 mg, but the intake of larger quantities could be required if poor absorption occurs (Sollman, 1957). The iron standard given for drinking or domestic water (0.3 mg/l) is meant to prevent objectionable tastes or laundry staining and is of aesthetic rather than of toxicological significance, although iron at high concentrations is toxic to livestock by interfering with the metabolism of phosphorus (NAS, 1974). For some industries, iron concentrates in process waters of lower than 0.3 mg/l are required or desirable. Examples of industries where water low in iron concentrations is required include: high pressure boiler feed waters, scouring, bleaching and dyeing of textiles, certain types of paper production, chemicals, food processing, and leather finishing industries (U.S. EPA, 1976).

MANGANESE

Manganese does not occur in nature as a distinct metal but occurs consistently in various salts and minerals (U.S. EPA, 1976). It is also frequently associated with iron compounds. Principal manganese-containing substances include manganese dioxide, pyrolusite, manganese carbonate (rhodocrosite) and manganese silicates (rhodonite).

Manganese is a minor constituent of igneous rocks (937 mg/kg; Matthess, 1982), and its rarity in seawater and occurrence in resistates, hydrolysates, and precipitates reflects low geochemical mobility. The element, however, causes considerable difficulties and is a troublesome groundwater constituent.

Like iron, manganese occurs in the divalent and trivalent form and its chlorides, nitrates, and sulfates are highly soluble in water (McKee and Wolf, 1972). However, its oxides, carbonates, and hydroxides are only sparingly soluble. In groundwater areas that are subject to reducing conditions, manganese is leached from the soil and occurs in high concentrations in the water. Manganese in such waters accompanies iron, although they normally occur together.

Manganese in igneous rock minerals occurs usually in the reduced form (Mn^{++}), substituting for some other divalent ion of similar size (Hem, 1970). However, the most common forms of manganese in rocks and soils include oxides and hydroxides in which the oxidation state of the element is +2, +3, or +4.

Manganese is also an essential element in plant metabolism, and the organic circulation of the element influences its occurrence in natural water (Hem, 1970). Various species of trees are very effective accumulators of manganese (Bloss and Steiner, 1960; Ljunggren, 1951), and certain aquatic plants also accumulate high levels of the element (Oborn, 1964). After leaf fall or when plants die and decay, manganese is liberated again. Microorganisms play an important role in the oxidation and reduction of manganese (Matthess, 1982).

Manganese is detectable in minor amounts in most groundwaters. In addition, under reducing conditions, element concentrations above 1 mg/l are relatively rare, although values of 0.05 mg/l have adverse effects on the potability of water (Matthess, 1982). The 1962 Drinking Water Standards (US PHS, 1962) recommended a limit of 0.05 mg/l for manganese. The 1958 WHO International Standards (WHO, 1958) prescribe a "permissible limit" of 0.1 mg/l and an excessive limit 0.5 mg/l, but no maximum limit is given. The 1961 WHO European Standards (WHO, 1961) have a recommended limit of 0.1 mg/l. These previously listed limits were established on the basis of aesthetic and economic considerations rather than specific physiological hazards (McKee and Wolf, 1972).

Diets deficient in manganese result in impaired or abnormal growth, symptoms of central nervous system disturbance, anemia, and possible interference with reproductive functions (Browning, 1961; Rothstein, 1953). The daily intake of manganese required for a normal human diet is 10 mg. The element is absorbed slightly and accumulates in liver and kidney tissues.

Manganese is undesirable in water supply systems because it causes unpleasant tastes, stains laundry and plumbing fixtures, and fosters the growth of certain microorganisms in distribution systems (McKee and Wolf, 1972; Anonymous, 1958; Connelley, 1958; Griffin, 1958, 1960; Myers, 1961).

However, most industrial users of water high in manganese can operate successfully where the criterion proposed for public water supplies is observed (U.S. EPA, 1976). Examples of industrial tolerance of manganese in water have been summarized for industries such as dyeing, milk processing, paper, textiles, photography, and plastics (McKee and Wolf, 1963, 1972).

SULFATE AND SULFUR SPECIES

The element sulfur when dissolved in water occurs in the fully oxidized (S^{+6}) state complexed with oxygen as the anion sulfate (Hem, 1959, 1970). Some reactions of sulfur (i.e., such as reduction) take place in natural systems unless certain species of bacteria, together with a suitable food or energy supply, are present (Matthess, 1982). One characteristic reaction within the groundwater zone is microbial catalysis of the reduction of sulfate by members of the obligate anaerobic genus *Desulfovibrio*, which extract the energy and hydrogen necessary for life from organic substances. The *Desulfovibrio* group, represented by *D. desulfuricans*, is adapted to the specific salt content and temperature of the aquifer.

Pyrite crystals occur in sedimentary rock and constitute a source of both ferrous iron and sulfate in groundwater (Hem, 1970). Pyrite is commonly associated with deposits such as coal, which was deposited under strongly reducing conditions.

Sulfate occurs in certain igneous-rock minerals of the feldspathoid group, but the most extensive and important occurrences are in evaporative sediments (Hem, 1970). Calcium sulfate as gypsum or as an anhydrite (i.e., which contains no water of crystallizations) makes up a considerable part of evaporite-rock sequences. Sulfates also occur in oxidized states of organic matter in the sulfur cycle and in turn serve as sources of energy to split bacteria (McKee and Wolf, 1972). The chemical also is discharged in industrial wastes, e.g., from tanneries, paper mills, textile mills and other operations that use sulfates or sulfuric acids.

The 1962 Drinking Water Standards (USPHS, 1962) recommends that sulfates not exceed 250 mg/l. This limit was not set on taste or physiological effects, but rather because of a laxative action occurring in new users.

CHLORIDES

Chloride is present in rock types in lower concentrations than any other major constituents of natural water (Hem, 1959, 1970). Chloride-bearing minerals occurring in igneous rock include feldspathoid sodalite, and the phosphate mineral, apatite. Johns and Huang (1967) summarized the data on chlorine content of rocks. However, igneous rocks do not yield high concentrations of chloride to normally circulating natural water (Hem, 1970).

Chloride may also be present in resistates as the result of inclusion of connate brine and is expected to occur in incompletely leached deposits laid down in the sea or in closed drainage basins. Porous rocks when submerged in the sea become impregnated with soluble salts in which chloride plays a major function (Hem, 1970). Fine-grained marine shale also retains chloride for long periods. It is present either as sodium chloride crystals or as a solution of sodium and chloride ions in this shale.

Chlorides eventually concentrate in marine and terrestrial evaporite deposits (Matthess, 1982). The chloride ion also occurs in terrestrial dust, volcanic emanations, and as human-caused airborne pollutants in the atmosphere. It also enters into the hydrologic cycle via chloride containing liquid and solid waste materials, in chloride-containing fertilizers, and highway salt. Chloride is also the dominant anion in acid groundwaters associated with active or recent volcanism (Matthess, 1982; White et al., 1963; McKee and

Wolf, 1972). The element occurs in practically all natural waters (McKee and Wolf, 1972). Chloride may be derived from: seawater contamination of underground supplies, salts spread on fields for agricultural purposes, human or animal sewage, or from industrial effluents (e.g., paperworks, galvanizing plants, water softening plants, oil wells, and petroleum refineries).

Chlorides in drinking water generally are not harmful to human beings until high concentrations are reached. However, they may be injurious to people suffering from diseases of the heart or kidneys (McKee and Wolf, 1972). Restrictions on chloride concentrations in drinking water are based on palatability requirements rather than on health. They may impart a salty taste to the water at concentrations of 100 mg/l, although in some waters 700 mg/l is not noticeable (Anonymous, 1947, 1950).

Chloride tolerance by humans varies with climatic conditions and level of exertion. Chlorides lost through perspiration may be replaced by the element in either the diet or drinking water (McKee and Wolf, 1972). Chloride limits, both those in use as standards and those recommended for use, vary over a wide range. It is believed that this relates significantly to the chloride source. If levels at the source are high, pollution may be occurring and chloride levels should be further investigated as a cause (McKee and Wolf, 1972).

The USPHS Drinking Water Standards (USPHS, 1962) recommended that chloride does not exceed 250 mg/l.

FLUORIDE

Fluorine, like chlorine, is a member of the halogen group, but its behavior in rocks and in the weathering process is different. Fluorides are low in solubility and amounts present in ordinary waters are limited (Hem, 1959). Fluoride frequently occurs in igneous and metamorphic rocks as a component of amphiboles (e.g., hornblende) and of the micas (Hem, 1959, 1970). Alkalic rocks and obsidian are high in fluoride when compared to other igneous rocks (Rankama and Sahama, 1950). Fluorine is a biologically important constituent of teeth and bones and is present in industrial flue gases (Matthess, 1982). Fluoride forms strong soluble complexes with aluminum, beryllium, and ferric ion, and in the presence of boron, mixed fluoride-hydroxide complexes occur (Matthess, 1982).

The concentration of fluoride in natural water which has a total dissolved solids content of less than 1,000 mg/l is usually less than 1 mg/l (Hem, 1970), although concentrations of 50 mg/l have been reported in solutions that are not potable. Fluorides are also used as insecticides, for disinfecting brewery apparatus, as a flux in the manufacture of steel, preservation of wood and mucilages, manufacturing of glass and enamels, and for water treatment (McKee and Wolf, 1972).

Fluorides at high levels are toxic to humans. Doses of 250 to 450 mg cause severe illness and 4.0 g can cause death. Numerous articles occur in the literature that describe the effects of high fluoride-bearing waters on dental enamel of children. Some articles pertain to excessive skeletal damage (McKee and Wolf, 1972). However, water containing less than 1.0 mg/l of fluoride seldom causes mottled enamel in children. Concentrations less than 3 or 4 mg/l are not likely to cause fluorosis and skeletal effects in adults (Vaughn et al., 1958; Anonymous, 1950b; Hinman, 1938; Kehoe, 1944; Smith, 1935; Smith et al., 1937).

NITRATES

Two gases, molecular nitrogen and nitrous oxide, and five forms of non-gaseous, combined nitrogen, viz., amino and amide groups, ammonium, nitrite and nitrate, are important in the nitrogen cycle (U.S. EPA, 1976). Amino and amide groups occur in soil organic matter and are an important constituent of plant and animal protein. Nitrate ions are formed from nitrate or ammonium ions by specific microorganisms that occur in soil, water, sewage, and the digestive tract of organisms (U.S. EPA, 1976). Nitrate ions are formed by the oxidation of ammonium ions by soil or water microorganisms (i.e., nitrite is another intermediate product of this nitrification process).

Growing plants assimilate nitrate or ammonium ions and convert them to protein. This is done by a process known as denitrification. Nitrate-containing soils become anaerobic during this process and the conversions to nitrite, molecular nitrogen, or nitrous oxide occur (U.S. EPA, 1976).

Nitrogen entry into water bodies occurs principally through municipal and industrial waste waters, although septic tanks, and feedlot discharges also contribute significantly (U.S. EPA, 1976). Other diffuse sources of nitrogen include farm-site fertilizers, animal wastes, lawn fertilizers, leachates from waste disposal in sanitary landfills, atmospheric fallout, nitric oxide, and nitrite discharges from automobile exhausts. Losses from natural sources such as mineralization of soil organic matter also form nitrate sources.

Nitrogen in small amounts is present in rocks, but the element is concentrated in soil and biological material (Hem, 1959, 1970). The greater portion of nitrogen on earth occurs uncombined throughout the atmosphere. Nitrogen compounds are also added to rainwater and its ammonia content during precipitation is 0.01–1 mg/l (Matthess, 1982; Junge, 1963). Organically bound nitrogen is then oxidized via amino acids and ammonia to nitrite and these compounds are transformed by biological process (Matthess, 1982).

High intake of nitrates by warm-blooded animals, particularly those under three months of age, constitute a hazard, primarily when conditions occur that are favorable to their reduction to nitrite (U.S. EPA, 1976). Nitrate is reduced to nitrite in the gastrointestinal tract. The nitrites then reach the bloodstream and react directly with hemoglobin to produce methemoglobin, which causes impairment of oxygen transport. Fatal poisonings in infants have occurred following ingestion of well waters containing nitrate at concentrations greater than 10 mg/l nitrate nitrogen (Stewart, 1967). Over 2,000 cases of infant methemoglobinemia have been reported in Europe and North America since 1945; with 7 to 8 percent of the affected infants dying (Walton, 1951; U.S. EPA, 1976). Susceptibility of infants to methemoglobinemia appear to be related to factors such as nitrate concentration, occurrence of enteric bacteria, and the lower acidity which exists in the digestive systems of young mammals (U.S. EPA, 1976). Consequently, waters with nitrite nitrogen concentrations over 1 mg/l should not be used for infant feeding (U.S. EPA, 1976).

PHOSPHATES

Phosphate in igneous rocks is largely concentrated in apatie. Apatie is a complex phosphate of calcium which contains fluoride, chloride, and hydroxyl ions in varying proportions (Hem, 1959). Weathering of these rocks releases calcium phosphate. Calcium phosphate is soluble in water containing carbon dioxide.

Phosphate is made available to water bodies as a by-product of several kinds of cultural applications of the elements which occur through activities of man (Hem, 1970). Phosphates consequently occur in surface or groundwaters as a result of: leaching from minerals or ores in natural processes, degradation of cleaning products, industrial wastes, and as a major element of municipal sewage (McKee and Wolf, 1972).

Phosphate is important to plant and animal metabolism and occurs in their waste products (Matthess, 1982). The use of sodium phosphate in detergents in many instances increases the supply of phosphorus in natural waters. Phosphates enter waterways from different sources, and the human body excretes one pound per year of phosphorus (U.S. EPA, 1976). The use of phosphate detergents increases the per capita contribution to about 3½ pounds per year of phosphorus. Some industries, such as potato processing, have wastewaters high in phosphates. Tilled, forested, fallow, and urban land can contribute varying amounts of phosphorus to drainages and water courses. Cattle feedlots, concentrations of domestic duck or wild duck populations, tree leaves, and fallout from the atmosphere, also contribute phosphorus (U.S. EPA, 1976).

In summary, evidence indicates the following (U.S. EPA, 1976):

- High phosphorus concentrations are associated with accelerated eutrophication of waters.
- Aquatic plant problems develop in reservoirs and

other standing waters when phosphorus levels become high.

- Reservoirs and lakes that become sinks for phosphates entering through influent streams store a portion of the element within their consolidated sediments and serve as a phosphate depository.

Hydrogen Ion Activity (pH)

A measure of the hydrogen ion activity in a water sample is known as pH. It is mathematically related to hydrogen ion activity according to the expression:

$$pH = - \log (H^+)$$

where (H^+) is the hydrogen ion activity.

The hydrogen ion activity of natural waters is a measure of acid-base equilibrium achieved by the various dissolved compounds, salts, and gases (U.S. EPA, 1976). The principal system regulating pH in natural waters is the carbonate system, which is composed of carbon dioxide, carbonic acid, bicarbonate ion, and carbonate ions. The interactions and kinetics of this system are described by Stumm and Morgan (1970).

The pH is an important regulator in chemical and biological systems of natural waters. The degree of dissociation of weak acids or bases is also affected by changes in pH. This effect and change is important because the toxicity of many compounds is affected by the degree of association (U.S. EPA, 1976).

The hydrogen ion concentration of a raw water source for a domestic water supply system is important in that it can affect taste, corrosivity, efficiency of chlorination, treatment processes (e.g., coagulation), and industrial applications (McKee and Wolf, 1972). The United States Public Health Service Drinking Water Standards of 1962 (USPHS, 1962) recommended no limits in reference to pH in domestic waters. However, natural waters do not dissolve lead if their pH value is above 8.0 (Hall, 1937), and it acquires a "sour" taste at pH 3.9 or below (Rohlich and Sarles, 1947; from McKee and Wolf, 1972). As a historical note, during Prohibition, waterworks operators received numerous complaints from customers who preferred water with a pH below 7.0 for home brewing (Collins, 1937).

Groundwater with a pH less than 6.0 is also potentially dangerous to concrete through corrosive properties (Ley, 1941). In this instance, the corrosion of concrete apparently is not a function of pH alone but varies with the content of hardness and alkalinity constituents, with free carbon dioxide being particularly detrimental. Low pH also favors the formation of carbonic acid from bicarbonates (Hammerton, 1945; Anonymous, 1938).

Theoretical concepts and various practical aspects of pH determinations have been discussed at length in the literature. Bates (1964) and Hem (1959) have prepared extensive summaries of the topic.

ALKALINITY

Alkalinity is the sum total of components in water that tend to elevate its pH above a value of 4.5 (McKee and Wolf, 1972). It can be expressed as mg/l of calcium carbonate, a measurement of the buffering capacity of the water (i.e., since pH has a direct effect on organisms on the toxicity of certain contaminants in water, the buffering capacity of a system is important). Alkalinity is not a specific contaminating substance, but rather includes a combined effect of several substances and conditions (McKee and Wolf, 1972). Examples of common materials in natural waters that relate to alkalinity, include carbonates, bicarbonates, phosphates, and hydroxides. It is also caused, to a lesser extent, by borates, silicates, phosphates, and organic substances. For the most part, alkalinity is produced by anions or molecular species of weak acids which are not fully dissociated above a pH of 4.5 (Hem, 1970).

In itself alkalinity is not considered detrimental to humans but is generally associated with high pH values, hardness, and excessive dissolved solids, all of which may be deleterious (McKee and Wolf, 1972). Natural waters, specifically in the southwestern U.S., are highly alkaline, while those in western Washington or New England are low in alkalinity (McKee and Wolf, 1972).

The best waters for the support of diversified aquatic life are those with a total alkalinity of 100 to 120 mg/l or a pH value between 7-8 (Huet, 1948; Barrett, 1952). Alkalinity in this instance serves as a buffer to help prevent sudden change in pH, which can cause death to aquatic organisms.

ACIDITY

"Acidity" as applied to water, is defined by the American Society for Testing Materials (1970) as "the quantitative capacity of aqueous media to react with hydroxyl ions." The definition of alkalinity given by the same reference substitutes the word "hydrogen" for "hydroxyl." In contrast, the alkalinity determination can be defined for all waters as a determination of the carbonate and bicarbonate concentrations. Major constituents of natural water react with hydrogen (H^+) and hydroxide (OH^-) ions and can be defined as acidic or alkaline.

In natural systems where the production of H^+ is rapid in comparison to the rate at which it reacts with minerals available, the pH remains more acidic than the range reached at equilibrium in carbonate systems (Hem, 1970).

Trace Metals in Natural Water

Trace metals (e.g., zinc, cadmium, copper, silver, lead, chromium, barium, molybdenum, aluminum, beryllium, nickel, cobalt, vanadium) may be present in natural groundwater or surface water (NAS, 1977). The sources of many trace metals are associated with natural processes, although many of them are due to man's activities.

Trace metal sorption by coagulated humic acid and peat materials can be described macroscopically either as a complexation or a cation exchange phenomenon (Sposito, 1986). Model equilibrium constants or selectivity coefficients for trace metal sorption by H-form humic materials are special cases of two general models which do not depend on the sorption mechanism. Sorption mechanisms can be elucidated by spectroscopic techniques, both optical and magnetic resonance, but no information about mechanism can be inferred from equilibrium constants (Sposito, 1986).

The two most important natural processes that contribute trace metals to natural water include chemical weathering and soil leaching (NAS, 1977). The factors affecting this chemical weathering and soil leaching process include solubility, pH, adsorption, hydration, coprecipitation, and colloidal dispersion (NAS, 1977).

Decaying vegetation also affects concentrations of trace metals in water and many plants are known to store elements selectively (NAS, 1977). Rainwater as it falls through the atmosphere and moves through soil also picks up trace metals that eventually end up in groundwater.

The literature on trace metals in natural water is voluminous, specifically on the chemical analyses of certain elements. The National Environmental Research Center of the U.S. EPA at Cincinnati, Ohio, has also published many methods of chemical analyses of water and wastes (U.S. EPA, 1973), although atomic-absorption methods are suggested in most instances. Publications such as *Standard Methods for the Examination of Water and Wastewater* (American Public Health Association, 1965, 1976, 1986) and the *Annual Book of ASTM Standards* (American Society for Testing Materials, 1970) recommended procedures for selection of sampling sites, frequency of sampling, sampling equipment, sample preservation, laboratory equipment, laboratory techniques, accuracy and precision of analyses, and reporting of data. Numerous symposium volumes and handbooks on water analyses have also summarized the state of the art for trace metals in aqueous solution. These reviews are described in Hume (1967), Hemphill (1972), and Cosgrove and Brocco (1973).

Recent developments concerning analytical methods used for trace metal analyses in waters are reviewed bi-annually in *Analytical Chemistry* and annually in the *Journal of Water Pollution Control Federation* (Minear, 1975). For general information on atomic-absorption analysis, books by Elwell and Gidley (1966), Slavin (1968), Ramirez-Muniz

(1968), Price (1972), and Robinson (1975), should be consulted.

Retention of inorganics is also described in the proceedings of a specialized seminar of the International Association on Water Pollution Research and Control (*Water Science and Technology, Vol. 17,* No. 9, 1985). The mobilization of heavy metals in river sediment by nitrilotriacetic acid (NTA) (Loch and Lagas, 1984), distribution of metal pollution in groundwater determined from sump sludges in wells (Hoen and Gunten, 1984), and trace metal speciation in a soil profile irrigated with waste waters were also presented (Mattigod et al., 1984).

Organic Solutes in Natural Water

Previously, it was assumed that untreated groundwater could be used as an uncontaminated reference against which the contamination of surface-derived drinking water could be assessed (Barbash and Roberts, 1986; NAS, 1977; Council of Environmental Quality, 1981; Gottleib, 1985). However, the discovery in 1977 of high concentrations of more than eight synthetic compounds in drinking water led to the closure of sixteen private wells in the town of Gray, Maine. Since then, at least thirty-three toxic organic chemicals have been detected in drinking water in thirty-four states (Barbash and Roberts, 1986).

The severity and extent of the contamination of groundwater-derived drinking water by volatile organic compounds (VOC's) could therefore be large in the United States (Coniglio, 1982). A broad range of chemical contaminants have been, or are likely to be, detected (National Resources Defense Council, Inc., 1982). These would include synthetic and petrogenic organic compounds, toxic metals, and radionuclides. In addition, many VOC's have shown evidence of animal or human carcinogenicity, mutagenicity, or teratogenicity (NAS, 1977). VOC's are also quite persistent in groundwater because of their low reactivity and may be transported for long distances. They are also used in both domestic and industrial applications that can result in subsurface disposal (e.g., degreasing operations, septic system cleaning, pest and weed control, dry cleaning, fumigation) (Westrick et al., 1983).

In the United States, approximately 20 million single housing units have on-site liquid waste disposal systems. Oftentimes, to clean out a filled septic system, typically, a gallon or less of cleaning fluid (e.g., trichloroethylene, methylene chloride, benzene) is flushed down the toilet once every one to two years (Barbash and Roberts, 1986). Septic discharges therefore do represent one of the most common sources of toxic organic contamination in the United States.

Organic substances dissolved in groundwater originate either from soil zones or aquifer materials. The substances include humic acids, pectins, hydrocarbons, simple fatty acids, tannic acids, and their degradation products (Csajaghy, 1960; from Matthess, 1982). Polycylic aromatics which are carcinogens, are also detectable in natural groundwater at concentrations of 1–10 mg/l to depths of fifty meters or more (Kruse, 1965; from Matthess, 1982). Other polycyclic hydrocarbons occur as natural products of metabolism and also in manmade pollution. Because they are soluble substances they reach the water table in appreciable amounts under the same circumstances in which bacteria and viruses reach the same area (Borneff and Knerr, 1960; from Matthess, 1982). Matthess (1982) describes organic substances in groundwater, specifically the hydrocarbons methane and ethane.

The organic contaminants identified in drinking water constitute a small percentage of the total organic matter present in water (U.S. EPA, 1977). Although approximately 90 percent of the volatile organic compounds in drinking water have been identified and quantified, these represent no more than 10 percent of the total organic material of the nonvolatile organic compounds. However, there is agreement that VOCs in groundwater-derived water supplies represent a significant hazard to public health. The severity of this problem is likely to get worse before it gets better. Because of the long residence times, the full impact of the effects of VOCs on water supplies has yet to be felt. Even if present efforts to achieve source control and treatment are successful, the legacy of past carelessness will prevail for many years.

Water Science and Technology (Vol. 17, No. 9, 1985) in discussing degradation, retention, and dispersion of pollutants in groundwater, also included papers on: sorption of hydrophobic trace organic compounds in groundwater systems (Schwarzenbach and Westfall, 1985); retardation of toxic chemicals in a contaminated outwash aquifer (Patterson et al., 1984); and biodegradation of methanol and tertiary butyl alcohol in subsurface systems (Novak et al., 1984).

Trihalomethanes in Groundwater

Aqueous chlorine reacts with certain precursors in water to form trihalomethanes which are suspected carcinogens–mutagens. In a study conducted by Varma et al. (1984), groundwater supplies in six small towns in Maryland were evaluated. Samples of both raw and finished water were collected and analyzed for THMs and for THM formation potential at ambient pH and at pH 9. The results of THM formation potential generally showed a slight increase in formation potential at higher pHs. The results indicate that more attention needs to be given to monitoring groundwater supplies in small towns.

Toxicological aspects of organic compounds are reviewed by Kool et al. (1982) in relationship to:

(1) The presence of organic compounds in drinking water and possible health effects

(2) Epidemiological studies on the health effects of water-borne diseases caused by organic constituents

(3) Introduction and removal of toxic organics during treatment process

(4) Toxicity testing of organic mixtures concentrated from drinking water

(5) The significance of organic constituents in drinking water for human health

Microbiological Contamination of Groundwater

Historically, groundwater has been a source of clean, high quality drinking water, needing no treatment. In contrast, surface water has required some level of treatment because of increasing human development in the forms of industry, agriculture, aquaculture, recreation, and sewage disposal. The purity of groundwater has also led to reliance on this source for suppplying water for drinking purposes. However, between 1946–1974, more than 50 percent of water-borne disease outbreaks in the United States were attributed to contaminated groundwater sources (Allen and Geldreich, 1975). One of the important diseases spread via groundwater was hepatitis A. In a review of forty-eight hepatitis outbreaks occurring worldwide involving 31,357 people, groundwater was implicated in twenty-one, and involved 1,387 people (Taylor et al., 1966).

The land disposal of sewage effluent and sludge prompted the examination of chemical and microbiological effects upon groundwater. Also, as sewage treatment methods improve, production of sludges increases and results in greater impacts on land and aquifers. The use of groundwater as recharge with wastewater is also of concern.

Volumes of wastewater produced in the year 2000 that might be used as recharge are expected to be 10–15 times higher than those produced in 1970 (Ilvovitch, 1977). Consequently, we are dealing with problems whose solutions should not wait until the future. The protection of our groundwater resources is important, because our reliance on land disposal of wastes has increased and it is now an alternative to discharging them directly into surface water.

Microorganisms Indigenous to Groundwater

Groundwater is low in organic matter (i.e., that present in an unavailable form). This, combined with filtration in the unsaturated zone, precludes many organisms from forming, except for a few chemolithotropic bacteria (Hvid, 1955; from U.S. EPA, 1984a). The ratio of chemoorganotrophs to chemolithotrophs is related to the organic content of soil, which decreases with depth. Microorganisms decrease with depth of the soil and organic matter and numbers of chemoorganotrophs are low when compared to chemolithotrophs which rise in magnitude. Samples taken from deep borings usually are sterile or contain chemolithotrophic bacteria.

Methane occurrence in groundwater shows activity of methane-generating bacteria such as *Methanobacterium*, which reduce carbon dioxide and organic acids to methane. Such bacteria cause problems because methane is a nutrient source for slime bacteria which grow in filters, along pipes, and in containers (Volker et al., 1977). Additionally, sulfates, if present in groundwater, are reduced to hydrogen sulfide by bacterial organisms such as *Desulfovibrio* and render water unacceptable for drinking.

Iron and manganese bacteria may also be present in groundwater. These bacteria obtain energy by oxidation of ferrous and manganese compounds. Another bacteria, *Hydrogenomonas,* which can oxidize molecular hydrogen to obtain energy, has also been recovered from groundwater (Hvid, 1955; from U.S. EPA, 1984a).

Boring samples taken from deep areas contain few chemoorganotrophic bacteria unless some mixing with surface water has resulted in availability of organic material. Anaerobic microorganisms (e.g., *Actinomycetes, Clostridia,* anaerobic gram-positive non-motile cocci) in samples from borings have been reported by Danish investigators (Hvid, 1955; from U.S. EPA, 1984a). These investigators also have described non-sporeforming, motile, gram-negative, glucose-negative, and gelatin-positive bacteria in groundwater. Bacteria such as these are not detected by conventional methods, and propagate in stagnant areas.

Bacteria also propagate in potable water that has been stored. Investigators in Norway stored groundwater with a potassium permanganate value of 1.5 at 22°C for three to thirteen days, after which the original bacterial numerals of 2 per ml swelled to 100,000 per ml. These marginal levels of organic matter content, however, would be highly selective for the less fastidious bacteria (U.S. EPA, 1984a).

Fungi, of which 200 genera have been isolated from soil, are more prevalent than bacteria in soil. However, little is known about them in groundwater (U.S. EPA, 1984a). Fungi, pathogenic to humans, occur in soil and since they are aerophilic they do not grow under anaerobic conditions such as would be expected in water-logged soil.

Protozoa (e.g., flagellates, amoebae, and ciliates) appear in the upper portion of the unsaturated zone in numbers varying from 1000–1,000,000 per gram of soil. Because of their size, however, they become entrapped within soil matrixes and do not enter groundwater.

The study of viruses in water is fairly recent (Slade, 1985). Those responsible for the quality of drinking water have become uncomfortably aware of the existence of a large group of pathogens which may be present in their water. However, in general, viruses which have been collected from groundwater sources occur at lower concentrations than in surface waters, although significant numbers of viruses have been identified in the absence of bacterial indicator organisms. It appears as if the bacterial indicators fail because of bacteria's different survival rates among viruses and their different movement rates through underground strata. Much work is still needed on viruses.

A review of the literature on monitoring the viral content of drinking water reveals that the key question of the health significance of low concentrations has not been answered (Bitton et al., 1986). The findings point to the possible need for revising the classical water quality indicators such as turbidity, free-chlorine residual, and coliform content.

The water virology issue of *Water Science and Technology* (Vol. 17, No. 10, 1985) summarizes many aspects of the subject, including: detection of rotavirus in treated drinking water (Keswick et al., 1984); rotavirus survival in raw and treated waters and its health implications (Satter et al., 1984); genomic analysis of RNA viruses isolated from water (Hodgkiss and Moodie, 1984); detection of hepatitis A virus (HAV) in drinking water (Sobsey et al., 1984); analysis of tap water for viruses: results of a survey (Guttmann-Bass and Fattal, 1984); viruses and bacteria in a chalk well (Slade, 1984); detection and stability of enteric viruses in sludge, soil, and groundwater (Jorgensen and Lund, 1984); and others.

Microorganisms and Organic Substances in Groundwater

Undisturbed groundwater bears organisms and chemicals indigenous to that environment. However, microbial contaminants can enter the aquifer directly through openings (e.g., crevices, fractures) in bedrock or porous subsoil. They may percolate through shallow soils or enter the aquifer through various infiltration practices. Microbial contaminants may also enter groundwater areas via injection wells used for disposal of industrial or municipal wastewaters. Once these contaminants enter groundwater, they can travel great distances, especially if many crevices and fissures occur in the soil.

Organic substances in groundwater, however, depend on the presence of saprophytic microbes and dissolved oxygen (U.S. EPA, 1984a). Organic substances in the presence of sufficient oxygen will oxidize or mineralize to carbon dioxide. Aerobic bio-oxidation of organic substances in an aquifer deplete the supply of dissolved oxygen and decrease the oxidation-reduction potential of the water. When this occurs, the microbial reduction of certain substances (e.g., ferric and tetra-valent manganese ions, ammonium, nitrate, methane, hydrogen sulfide, and other reduced sulfur compounds) becomes enhanced and continues to progress in deep aquifers (Schwisfurth and Ruf, 1976; from U.S. EPA, 1984a).

Inorganic or organic substances in deep groundwater areas subsequently can be oxidized when water is brought to the surface. Various microorganisms (e.g., *Sphaerotilus, Leptothris, Phragmidiothrix, Gallionella, Crenothrix*) then proliferate, producing abundant sheaths or slimes which are encrusted with oxidized materials (Schweisfurth, 1974;

Volker et al., 1977). Sheaths and slimes discolor and impart tastes and odors to water. They also cause considerable problems in treatment, especially the slimes produced that are encrusted with iron or manganese. Sand filters can be plugged by such slimes, and this in turn will interfere with the treatment and distribution of water for consumption.

Microorganisms and Survival in Soil and Groundwater

Survival of pathogens in soil and water is determined by evaluating the species of microorganism in question, temperature levels, amount of sunlight, rainfall, pH, soil type, moisture, organic content of soil, and the degree of microbial antagonism (i.e., production of antibiotics, such as actinomycin or streptomycin), predation, parasitism, and the infiltration rate. Different organisms also exhibit various survival rates in soil and water. Mycobacteria and spore-formers can survive longer in soil than thermo-tolerant coliforms. Climate as a factor in survival rates is more important in winter than summer. Contamination of groundwater is also more likely to occur during heavy rainfall periods.

Microorganisms present on soil surfaces are exposed to ultraviolet light (UV) and liable to dessication. They, therefore, are less likely to survive than those in moist soil. Less than 15 percent humidity also reduces the numbers of viruses and bacteria that survive longer in alkaline (e.g., limestone) than in acid (e.g., peat) soils (Moore et al., 1976). Waste stabilization ponds, on the opposite extreme, protect microorganisms against exposure to (UV) and increase survival rates. Organic compounds and their presence increase survival of microorganisms (i.e., bacteria can survive for extended periods in soil to which digested sludge has been applied; U.S. EPA, 1984a).

Viruses and their adsorption is influenced by pH, type and concentration of cations, organic matter, and soil particles (U.S. EPA, 1984a). Clay soils and consolidated sands facilitate virus adsorption. Adsorption increases with acidity of the soil. It may also be hindered or reversed with high concentrations of organic matter. Sites consisting of fractured rock or gravel and sand are unacceptable for waste disposal because viruses and other microorganisms can be moved long distances. Viruses also persist in groundwater which is clean compared to river or seawater where significant viral reductions occur. Investigators have recovered hepatitis A virus from a well situated 23 m from a cesspool (Neefe and Stokes, 1945).

Once bacteria have entered groundwater, they do not have to contend with (UV) light, although *Escherichia coli* has been described as surviving 100 days in subsoil water (Gerba et al., 1975). Fecal streptococci and thermo-tolerant coliforms have also been recovered at a depth of 33 m after being pumped into the aquifer (Jepsen, 1972).

From tests made of treatment lagoons about 1 percent of coliforms had survived and .001 percent were viable after two weeks (Klock, 1971). A large majority of pathogens seem to be eliminated before they reach the soil. Nevertheless, when soil beneath infiltration ponds became saturated with sewage effluent, thermo-tolerant coliforms were recovered at 4–8 m (Barrs, 1957). Therefore, if infiltration ponds are placed in sandy soil, organisms could penetrate into groundwater.

Endotoxins and Their Occurrence in Groundwater

Gram-positive bacteria predominate in soil and gram-negative bacteria do so in water (U.S. EPA, 1984a). Irrigation and disposal of sewage effluent into soils then contribute enormous numbers of gram-negative bacteria.

The antagonism that occurs between indigenous gram-positive soil bacteria and gram-negative water microorganisms results in the release of endotoxins, which are produced by gram-negative bacteria. Endotoxins that are formed then travel from unsaturated zone areas in soil into groundwater. The saturated zone situated beneath infiltration ponds in Denmark was sampled at depths of 1.9–16.5 m and observations showed that endotoxins occurred in amounts from 0.0001–0.000001 g/ml (Kristensen, 1978; from U.S. EPA, 1984a). The endotoxin concentrations observed were over 10,000 times the minimum dose (2 μg endotoxin per kg body weight) needed to produce clinically measurable effects by parenteral injection in humans (Mikkelsen, 1977; from U.S. EPA, 1984a). Since endotoxins are toxic upon entering the bloodstream, their presence in water supply systems poses special problems.

Evaluating Microbiological Quality in Groundwater

Groundwater supplies can be tested by the same microbiological techniques as for potable water (Allen and Geldreich, 1975). Though differences occur from one country to another, most of the testing includes coliforms and colony count evaluations. Tests made in the United States have demonstrated that total and thermo-tolerant coliforms occur in some community groundwater supplies (U.S. EPA, 1984a). Yet, the absence of coliform organisms does not ensure that pathogens are not present in groundwater (Lassen, 1972). Before improved indicator systems for assessing groundwater quality can be designed, however, more base-line studies are needed concerning their survival and movement in soil-groundwater interfaces.

Control of Undesirable Microorganisms in Groundwater

To divert tremendous volumes of wastewater from discharging into rivers and lakes, land disposal is being used as an alternative. Although the soil acts as a buffer against groundwater contamination, not all the answers are known. Besides studying altered conditions for survival, propagation, and transport of pathogens, water microbiologists should concentrate on oxidation and mineralization processes, specifically their effects on the growth of chemolithotrophic and chemoorganotrophic organisms (U.S. EPA, 1984a). Also needed is more research into potential groundwater contamination from biotoxins associated with land disposal activities. Better indicator systems are specifically needed that consider properties unique to soil-aquifer ecosystems.

Agricultural, industrial, and urban expansion are leading to many-fold increases in water consumption which will occur during the next 100 years. Increased demands for potable water will necessitate turning to nontraditional raw water sources. However, in order to avoid markedly lowering the water table in many areas, surface water resources will still have to be used to a greater extent than previously. At the same time, surface water is increasingly being used as a repository for chemical and biological wastes. This trend will put municipalities and even nations on a disaster course if corrective steps are not taken, particularly in areas where groundwater is also used adjacent to surface water.

THREE

On-Site Liquid Waste Disposal Systems

Description of Systems

Septic Tanks and Leachfields

In rural America, on-site liquid waste disposal systems came into use as electricity and inexpensive systems became available to these areas. More than 1 trillion gallons per year are discharged from septic systems to subsurface areas (CEQ, 1980; U.S. EPA, 1977) and according to the 1970 U.S. Census, nearly 30 percent or 19.5 million of all housing units in this country, are served by on-site private sewage treatment facilities (Anonymous, 1980; U.S. EPA, 1973). The number of septic tanks systems in use in this country indicates that approximately 32 percent of the population is still not served by primary, secondary, or advanced waste treatment facilities (Anonymous, 1980).

Data available from the Public Health Service and Federal Housing Administration (U.S. EPA, 1973) indicate that in addition to the 32 million people served by septic tanks in 1970, there are an unestimated number of systems at summer camps, cabins, Forest Service campgrounds, and other organized group locations, all which must depend upon subsurface disposal of their wastewater.

The use of septic systems in subdivisions following World War II was also an attractive alternative to land developers because their objective was to convert land profitably from single ownership of large parcels to multiple ownership of small plots. This would also alleviate the responsibility of the subdividers to furnish centralized water supply and liquid waste utilities (U.S. EPA, 1973). However, the result was that, through inadequate knowledge and control, septic tanks with badly constructed percolation systems were approved for small lots in urban subdivisions, sometimes for 2,500 or more houses (U.S. EPA, 1973). Consequently, the effect was to concentrate greater amounts of wastewater at local points of infiltration than had previously existed. A secondary effect of this was that the high density of these

systems increased the potential for polluting subsurface areas (DeWalle and Schaff, 1980; Hain and O'Brian, 1979).

Failure of on-site systems in many areas also occurred, but this knowledge was usually confined to householders, owners, or dispersed in files of many health departments (U.S. EPA, 1973). Only after a single agency, the FHA, became responsible for home loan insurance did it become known that the failure rate in many instances was 30 percent within two or three years after facilities had been built. Failure of these systems was attributed to clogging of soil and subsequent appearance of wastewater on the surface of the ground.

Craun (1981) reported that the overflow of septage from individual on-site systems was responsible for 41 percent of disease outbreaks and 66 percent of the illnesses were caused by contaminated groundwater. Some of the bacteria isolated from wastewater in these cases, included *Salmonella* (typhoid), *Shigella,* and enteropathic *Escherichia coli.*

Septic tank and soil absorption systems have a significant role in on-site domestic waste disposal strategies (Nightingale and McCormick, 1985; McClelland, 1977); however, their impact on water quality is variable. The rate of groundwater quality change is affected by factors that are site-specific in response to quantity and chemical quality of the septic tank effluent, soil properties, water table location, subsurface geology, climate, and vegetation and its management over and around the leachfield.

Cesspools

Cesspools are typically 5–6 foot diameter holes several feet below ground surface. The facilities were built to receive raw sewage directly from house discharge systems (e.g., bathroom, wash wastes, etc.). The larger solids settled to the bottom of the sump and were trapped inside while the liquid fraction was meant to seep out through openings in the sides and bottom of the hole. In such systems, separation of solids from liquid wastewater was poor and cesspools worked only in coarse or highly fissured materials. Thus, such systems caused raw sewage to move directly into groundwater areas causing the contamination potential of these systems to be high. While installation of cesspool systems are prohibited at present by health codes, their popularity in the past (i.e., primarily because of their ease of construction) has left hundreds of thousands currently in operation (U.S. EPA, 1977). Many illegal systems, also, are still being built today.

Control Measures for Septic Type Systems

The two categories which relate to control measures (U.S. EPA, 1973) include:

(1) Those which lead to restrictions on the use of septic type systems
(2) Those which are properly designed and installed in suitable soil

In the first category three situations are identified which in turn create a health hazard associated with unacceptable nuisance conditions, such as the presence of decomposing sewage effluent on the ground surface.

A more serious situation related to groundwater quality includes direct discharge of untreated septic tank, cesspool, or drainfield effluent indirectly into groundwater. This can occur through coarse gravel beds, fractured rocks, channels, or lava tubes. When this happens, the groundwater itself often carries for long distances sewage solids, bacteria, viruses, obnoxious tastes, odors, and other materials. Consequent danger to health and impairment of the aesthetic acceptability of water then affects those that drink the water.

Another groundwater contamination situation is somewhat similar in that it occurs when percolation systems are located below biologically active zones of the soil. Such systems cause problems because they are installed where the frost line is deeper than the biologically active zone or they are buried too deep, usually, because of improper construction techniques. In such situations, biodegradation of effluent from the system is confined to the partial degradation of organics under anaerobic conditions. The physical phenomenon of filtering and adsorption also continues to be effective, but breakdown products of organic matter enter the groundwater and move with it. Tastes and odors are introduced at this time and the organic fracture because of its biochemically unstableness, supports bacterial growth in areas where groundwater comes to the surface or is withdrawn through wells (U.S. EPA, 1973).

When acceptable soil conditions exist, septic tank units which will effectively treat the wastes generated by a household, still require periodic pumping and removal of solids collected in the storage tank. The septage material removed from the tank should be transported to a central treatment facility, dumped, lagooned, spread on land, or deposited at a landfill.

Privies

This form of human waste disposal is common in rural areas where lack of indoor pressurized water systems precludes the use of other systems. The pit privy is typically a small, shallow pit or trench which normally receives only human waste and paper. Properly constructed pit privies allow effective decomposition and treatment of human wastes. This of course means properly screened and constructed buildings for control of vermin or insects. The volume of water introduced to such facilities is relatively small, and problems associated with odors and disease carrying insects occur more frequently with pit privies than groundwater contamination.

Industrial or Commercial On-Site Liquid Waste Disposal Systems

Septic tanks and seepage systems at industrial facilities generally operate on the same principal as those serving single and multiple dwellings. The volume of effluent is greater but what is more important is the possibility of toxic wastes from each specific manufacturing process being incorporated with domestic wastes and ultimately migrating into aquifers that serve drinking water supply systems. These hazards can be magnified considerably depending on the toxicity of the wastes.

The density of industrial septic tanks is high and they number in the tens of thousands as compared to the millions of septic tanks and cesspools serving housing units across the country. Fortunately, many laws now require that they obtain discharge permits and dispose of their septage in approved locations.

Characteristics of Septage or Contaminants

According to Russelman and Turn (1978), septage from septic tanks is a biodegradable waste capable of affecting the environment through water and air pollution. Proper control of its disposal requires knowing the number of tanks installed and the rate of new installations; the quantity of septage being hauled and by whom; the generally used disposal practices and problems associated with them; and the regulatory framework controlling the disposal.

To assure a more complete role in disposal control programs, a community should establish fees consistent with administrative and treatment costs (i.e., whether disposal occurs at a wastewater municipal treatment plant and/or at a landfill site or collecting pond). This type of system would require initiation of a trip manifest system to establish and track the origin of each load. Each load would also be sampled to determine integrity of the contents in reference to addition and/or knowingly spiking of the septage with hazardous waste or toxic substances. Hazardous waste and/or toxic substances cannot be deposited at municipal wastewater facilities and/or landfill sites. They should go to an approved hazardous waste disposal site. Unfortunately, many areas of the U.S. do not make provisions for this.

Canter and Knox (1985) summarize much of the literature relative to the types and mechanisms of groundwater pollution from septic tank systems, and provide information on technical methodologies for evaluating the groundwater pollution potential of such systems. The book is organized into five chapters that include the following discussions: septic tank system regulation; design of septic tank systems; groundwater pollution from septic tank systems (i.e., including influent wastewater characteristics, septic tank treatment efficiency, septic tank effluent quality); and septic tank system modeling. The transport and fate of bacteria and viruses in soils and groundwater are addressed along with inorganic contaminants such as phosphorus, nitrogen, chlorides, metals, and contaminants such as cleaning agents and pesticides.

In general, there is much information available on septage (septic tank pumpings) characteristics (Eikum and Paulsrud, 1986). However, considering all the factors influencing septage quality (tank size and design, pumping frequency and procedure, climate and seasonal weather conditions, household frequency of use, etc.), as well as difficulties in obtaining representative samples, there is reason to expect a wide variation in the values given for different parameters (U.S. EPA, 1977). It is important from a design standpoint though, to be aware of this. Also, the initial source of the septage has a lot to do with its characteristics, whether it be from residential and/or commercial, industrial concerns.

Septage is a highly variable anaerobic slurry having large quantities of grit and grease, a highly offensive odor, the ability to foam, poor settling and dewatering abilities, high solids and organic content, and, quite often, an accumulation of heavy metals (U.S. EPA, 1977b). Septage disposal on land can include surface spreading and sub-sod injection, spray irrigation, trench and fill, sanitary landfills, and lagooning. All land disposal alternatives require analyses of soil characteristics, seasonal groundwater levels, neighboring land use, groundwater and surface water protection and monitoring, climate, and site protection, such as signs and fences.

Trip manifest ticket systems should also be initiated at the state or local level to assure the integrity of all loads hauled to any disposal site, whether it be a wastewater municipal plant, landfill, or acceptable pit. Hazardous or toxic substances should be disposed of in a properly approved hazardous waste disposal site at all times and not substituted or mixed with septage.

Sewage from individual homes consists of about 99.9 percent water (e.g., by weight), 0.02 to 0.03 percent suspended solids, and soluble organic and inorganic substances (Pelczar and Reid, 1965). Present in domestic sewage also are bacteria, viruses, and microorganisms from the digestive and respiratory tract, and the skin. Domestic sewage

composition is not always uniform and varies from day to day, hour to hour, and household to household. In 1977, wastewater volume into septic tank systems from typical households ranged from 40 to 45 gpd/person (U.S. EPA, 1977). The portions of this total from sources within the house relate to factors such as use of automatic washing machines and other water use habits of the occupants. U.S. EPA (1977) reported the following ranges:

Source	Percentage of Total Domestic Waste Load
Toilet	22–45
Laundry	4–26
Bath	18–37
Kitchen	6–13
Other	0–14

A wide variety of household contaminants can also enter into the wastewater flow from the major sources within a house. Organic chemical content of these wastes can come primarily from human wastes, soaps, detergents, food wastes, and/or indiscriminate dumping of poisons and or hazardous chemicals (U.S. EPA, 1977).

Domestic sewage, however, by itself is a complex mixture and its specific substances have yet to be fully identified, although certain collective characterizations have been made (U.S. EPA, 1977). The constituents which present the greatest threat to groundwater quality include:

- excessive concentrations of nitrates in drinking water which can produce bitter tastes and cause physiological distress, e.g., methemoglobinemia in infants
- high phosphate discharges which cause eutrophication in surface waters
- lead, tin, iron, copper, zinc, and manganese in high concentrations, from household pipes and human wastes so they become toxic
- high levels of sodium, chloride, sulfate, potassium, calcium and magnesium which create health hazards to most individuals (Symptoms can range from laxative effects to cardiovascular and renal diseases.)
- excessive BOD (Biochemical Oxygen Demand) in septic fluids discharged which depletes the dissolved oxygen content from surface waters or enhances the anaerobic conditions which can occur in groundwater areas
- fecal coliforms, the non-pathogenic species of bacteria, which indicate the potential presence of pathogenic microorganisms

The effects of septic system effluent on the quality of groundwater also depend upon (U.S. EPA, 1973):

- differences in chemical and physical characteristics between local groundwater and the water supply utilized by owners of on-site septic systems

- the actual type, range, and concentration of materials added to the water supply as a result of human use
- the changes in the wastewater which occur in the septic tank and disposal field system, including the biologically active soil through which it percolates
- the kind and amounts of material which move downward as the water percolates to the groundwater
- the rate and degree of mixing wastewater and groundwater which occurs

Water supply systems are less mineralized than groundwater derived directly from the same local soil horizon into which septic system wastewater has been discharged (U.S. EPA, 1973). However, in general, the water supply should essentially have the same spectrum of ions as occurs in groundwater (e.g., nitrates, sulfates, carbonates, chlorides, etc.), but to a lesser concentration.

Septic tank systems should involve only domestic wastewater, but unfortunately, material from other various household discharges include in some instances grease and organic refuse from the kitchen, and detergent wastes from cleansing activities in the household. On other occasions, knowingly or unknowingly, contaminants such as water softener regeneration brines, pesticides, drugs, cleaning solvents, oils, and unknown mixtures are disposed of in the septic system by the houseowner or family members.

Changes which have occurred in septic tank effluent soil absorption systems were reported by McGauhey and Krone (1967). The changes which occurred suggested that:

- Inert and organic particulate matter can be effectively removed by the first few centimeters of soil (i.e., clogging of the infiltrative surface occurs rather than the changes occurring in quality of the percolated water).
- Bacteria seem to behave like particulate matter as they move through various soils and are removed by straining, sedimentation, entrapment, and adsorption. They will also die in an unfavorable environment.
- Viruses can also be filtered by soil systems, principally by adsorption, and as effectively as bacteria.
- Of the 300 mg/l total dissolved solids added to water by domestic use, a great portion will appear as anions and cations normally occurring in water. An increase in the mineralization of groundwater is expected from septic streams, and phosphates can be effectively removed, whereas chlorides, nitrates, sulfates, and bicarbonates will move freely with percolating water.
- Synthetic detergents will be effectively destroyed by biodegradation processes, specifically in active aerobic soils.

The previously listed changes, however, only apply to septic tank effluent systems which discharge directly into the soil mantle of the earth in the biologically active zone. This is the zone where physical phenomena and aerobic stabilization of degradable organic material become effective (U.S. EPA, 1973).

Effluent from septic systems will increase total dissolved mineral solids in groundwater and introduce bacteria, viruses, and degradable organic compounds. If septic tank systems are spread adequately away from each other, this separation will tend to minimize the concentration of pollutants in any unit of groundwater receiving the wastes. However, in certain local situations, the effects may be difficult to measure. In areas such as Long Island, New York, and Fresno, California, where numerous on-site liquid waste disposal systems were installed in individual subdivisions, the effect of effluent from these systems has been readily detected (Perlmutter and Guerrerra, 1970; Schmidt, 1972; U.S. EPA, 1973).

Local and regional groundwater problems associated with the use of septic tank systems have also been identified with many regional problems in the eastern United States (Perkins, 1984). At present, there is no federal legislation directly governing small individual on-site septic tanks and their discharge. It has been and still continues to be the responsibility of state and local governments to assure that adequate disposal practices are followed.

Control Methods
for On-Site Liquid Waste Disposal Systems

Control of septic tank systems and their effluent on groundwater quality can be classified into three groups (U.S. EPA, 1973). These include:

(1) Areas where septic tank installations already occur
(2) Areas where new septic tank systems are to be installed
(3) Areas where alternatives to conventional septic tanks are not feasible

Of these situations, the first grouping is the most difficult to deal with because design of the system is beyond recall and degradation of groundwater may have already occurred. If system failure is involved, the situation in some instances, is largely self-curative (i.e., replacement systems are installed).

The inability of soils to filter effluent to groundwater eventually results in its appearance on the land surface, and if the amount of discharge from the subdivision is significant, conventional city sewerage or central treatment facilities are needed. However, if existing on-site systems are functioning satisfactorily, their contribution of salts to groundwater can be computed from initial water quality analyses of the main water supply. Information on what is added to the effluent from the household, however, is also necessary. Control programs to abate groundwater contamination would then involve an evaluation of the effluent and exclusion of toxic products. Systematic monitoring of the upper level of the aquifer and evaluation of its significance can also lead to recommendations that abate pollution and control its source. Other control procedures could also be applied, depending on results. Some of these could include:

- government entities requiring subdivisions subject to on-site liquid waste system failures (i.e., that are also damaging to groundwater quality) to enter into special assessment districts that have centralized sewage collection and treatment facilities planned for specific dates in the future
- government requiring householders to connect to municipal sewerage as such systems become available (i.e., close enough for hook-up, as defined by uniform plumbing codes, usually 200 feet or less to the particular building or household in question). As land development extends to populated areas beyond the initial subdivision, hook-up would be mandatory.
- government legislation, e.g., regulations, ordinances, policies, etc., prohibiting the home regeneration of water softeners, unless the effluent can be treated in separate septic tank systems and stored in holding tanks with periodic pumping

Where new on-site liquid waste disposal systems are proposed for any area, possible measures for control of such facilities include:

- requiring approval and evaluation of the site and design by competent soil scientists and qualified personnel, e.g., health scientists, environmental engineers, before disposal systems are accepted for any proposed subdivision or individual household development (Simple percolation tests and standard codes in many instances offer inadequate criteria for the design of a comprehensive septic system; replacement area sites for eventual drainfield failure should also be included.)
- constructing alternate effluent discharge systems by methods which do not compact drainfield areas, including:
  - limiting heavy equipment or weight upon the recognized leachfield surface(s)
  - use of adequate stone sizes in the leachfield backfill area to produce "clogging in depth" (McGauhey and Winneberger, 1964)
  - installation of observation well risers at the end of each leachfield
- operating on-site liquid waste leachfield systems effectively by:
  - alternately loading, resting, and reusing one-half the percolation or drainfield system, with the

change cycle determined by observing ponding in the system at the observation well, or alternating use of the system through established systematic time intervals (i.e., every 6 months, etc.)
  - loading the entire infiltrative surface of the system at each cycle as uniformly and simultaneously as possible
  - periodically inspecting and removing septage, scum, grease, etc., from the main septic tank annually, bi-annually, or as often as necessary
  - when drawing off septage material, half or less than half is kept, rather than pumping out the entire contents of tanks (This instills a faster seeding mechanism for new wastes so adequate bacterial growth and breakdown of products continue without significant interruption.)
- use of environmental health ordinances, zoning ordinances, subdivision ordinances, and other land management control and review processes to prevent on-site liquid waste disposal systems from being installed in unsuitable areas (i.e., soils too impervious to accept effluents, groundwater levels too high, too close to water supply wells, gradient too steep, soils impermeable, lot sizes too small, etc.)

In areas where alternatives to on-site liquid waste disposal systems are not feasible, other choices could include:

- limiting use of septic tank and leachfield systems to the growing season for vegetation
- permitting the use of septic tanks and leachfields if soils are suitable, and accepting the consequences in terms of systematic groundwater quality degradation
- permitting use of septic tank systems and leachfields, but restricting the materials which are discharged through them, i.e., prohibiting the installation and use of household water softening units, release of pesticides, oils, toxic materials, etc.
- permitting the use of septic tanks (i.e., holding tanks only) under specific conditions
- allowing no discharge to occur

Some of the previously listed alternatives can be applicable to installations such as forest camps, summer cottages, and summer camps in remote areas where evapotranspiration and plant growth consume large amounts of water and nutrients. Some of the other described alternatives are generally necessary when isolated dwellings occur on large plots of land remote from any central or municipal sewer lines.

Other alternatives listed include appropriate control measures if soils are suitable for infiltration of effluent and good design and operating procedures of the system are followed. Eventually, city sewers may be provided in the streets of a housing development and houseowners may be required to connect to the central sewer as it becomes available. Availability, in this sense could be defined via Uniform Plumbing Code requirements (UPC) or as defined in local resolutions and/or ordinances, i.e., 200 feet or less from a municipal sewer.

In specific instances or areas septic tank systems should not discharge liquid waste effluent into the soil, and a holding tank might be required. However, holding tanks should not be expected to be permanent solutions and pumping of them can be very expensive.

Effluent from Septic Tanks and Groundwater Contamination

Groundwater contamination in areas of Minnesota (1950–1959) where on-site liquid waste disposal systems persisted was attributed to the widespread use of individual water supply systems and septic leachfield drains (Woodward et al., 1961). By 1959, over 40 percent of the water supply systems of one Minneapolis suburb yielded groundwater of high chemical content. The report included results of a field study of ninety-eight private water systems in the city of Coon Rapids and 63,000 wells in the metropolitan Minneapolis area. Water samples were analyzed for nitrates, surfactants, coliforms, chloride contents, and 47 percent of the 63,000 wells were contaminated. Control methods to solve this problem included regulation of individual septic tank and leachfield construction, installation of a community water supply system, and eventual establishment of a central sewage disposal and collection system.

Polta (1969; from U.S. EPA, 1974) also presented a discussion of septic tanks as potential sources of nitrogen and phosphorus contamination to groundwater areas due to effluent discharge by means of tile fields and seepage pits. The extent of nitrogen flow in groundwater was affected by an adsorption-ion exchange phenomenon exhibited in soils, along with the action of nitrifying and denitrifying bacteria. Phosphorus was not seen as a serious contamination threat, but nitrogen in groundwater caused risks of eutrophication and methemoglobinemia.

Waltz (1972) reported on problems encountered in developing mountain homesites in the Rocky Mountains of central Colorado. These particular homesites required individual water supply wells and on-site liquid waste disposal systems, but the leachfield systems generally were not suited for use in mountainous terrain where soils were thin, missing, or at improper gradients. Contamination of groundwater from these improperly placed leachfield lines became a problem as the effluent entered bedrock fractures and then was able to travel large distances without purification occurring. Evaluation of geologic conditions in site selection areas of septic tanks, leach fields, and wells were recognized as a method of significantly decreasing water well contamination in this particular area.

Baker and Rawson (1972) in another study evaluated groundwater quality in the Toledo Bend Reservoir area of Texas. Twenty test wells were installed downgrade from septic tank and leachfield systems. In the spring of 1972, results showed that coliform density of shallow groundwater samples ranged from 0.0 to 1,800 colonies/ml. One sample per eighteen wells revealed coliform presence, and one sample per twelve of the test wells contained more than 100 coliform colonies/ml.

The effects of cleaning fluid on septic tanks has already been discussed under the section Organic Solutes in Natural Water (Chapter 2). Studies concerning the presence and level of certain trace organics in wastewater samples collected from a septic tank in an individual household, from a lift station, and from a wastewater lagoon near Regina, Canada, are presented by Viraraghavan and Hashem (1986). Out of eleven priority pollutants analyzed, six priority pollutants—chloroform, bromodichloromethane, toluene, methylene chloride, and tetrachloroethylene—were detected in the samples. Benzene and bromodichloromethane were dominant. However, the authors concluded that the trace organics in the septic tank effluent would not pose any significant risk because of the attenuation capacity of the soil and dilution factors. This could change, however, if the density of septic tanks per lot size would increase.

There is also still a need to further research the following (Viraraghavan and Hashem, 1986):

- intensive sampling of septic tank influent and effluent for other priority pollutants, such as phenol, 2,4,6-trichlorophenol, 2-chlorobenzene, 1,1,1-trichloroethane, naphthalene, diethyl phthalate, dimethyl phthalate

- laboratory column studies with different soil types to understand about the behavior of trace organics through soil

- field studies to monitor the presence and quantity of trace organics downstream of existing and experimental septic tile systems in different soil environments

- regular groundwater monitoring for trace organics in areas where there is a relatively high density of septic tank systems to assess the long-term effects of such systems on groundwater quality

Summary of On-Site Liquid Waste Disposal Systems

Septic tanks and their related effluent rank highest in total volume of wastewater discharged directly to groundwater sources and are frequently reported sources of contamination (U.S. EPA, 1977). However, as previously described, most problems relate to individual homesites or subdivisions where recycling can cause pathogenic organisms to survive, the health hazards from these systems being considered moderate, with relatively high concentrations of nitrate representing the principal concern.

Additionally, regional groundwater quality problems have been recognized in areas of great density where many systems occur. For example, across the United States, there are four counties (Nassau and Suffolk, New York; Dade, Florida; and Los Angeles, California) that have more than 100,000 housing units served by on-site septic tanks/leachfields and there are twenty-three counties with more than 50,000 (U.S. EPA, 1977).

Where the density of on-site disposal systems has created problems, collection of domestic wastewater by public sewers and treatment at a central facility has been a common alternative. Other alternatives (which are generally limited to special situations where natural conditions or restrictive codes rule out conventional septic tank systems, include aerobic treatment tanks, sand filters, flow reduction devices, evapotranspiration systems, and artificial soil mounds disposal systems (U.S. EPA, 1977).

Where sewer systems are not feasible, prevention of groundwater quality problems has been attempted by low density zoning at the local government level, although increased regulation of septic tank siting, construction and design is emerging at the state and local government levels. Many states and local municipalities now participate in septic tank permitting and regulations, and a large number also provide local zoning agencies and review boards sufficient data to make land-use planning and subdivision control decisions.

Technical Considerations of On-Site Liquid Waste Disposal Systems

The manual of septic tank practice (U.S. DHEW, 1967) describes a soil as being suitable for the absorption of septic tank effluent if it has an acceptable percolation rate, without interference from groundwater or impervious strata below the level of the absorption system. However, for a septic tank and leachfield system to be approved or certified by state or local environmental health agencies, several additional criteria must be met (U.S. EPA, 1977). These include a specified percolation rate as determined by an acceptable percolation test, and the maximum seasonal elevation of groundwater. There must also be a reasonable thickness of the soil between the lower end of the leachfield and the top of the water table. Normally it can be four feet in relatively permeable soil.

Keeping the septage below ground may have been sufficient under low density rural conditions for which these systems were originally designed, but their widespread use in numerous small lot sizes is what is largely responsible for the degradation of groundwater that has occurred. Appraisal of the actual or potential contamination of groundwater by on-site liquid waste disposal systems also requires a basic understanding of the groundwater system into which the effluent is being discharged (Holtzer, 1975). Groundwater

recharge areas and flow patterns which accept discharges from septic tank systems should also be understood. In addition, the quantity of groundwater recharge must be estimated to compute the degree of natural dilution of the effluent. The capability of the soil system to renovate effluent disposal should also be known.

Although these previously mentioned concepts surpass the many established septic tank and leachfield system siting criteria, they should be instituted if groundwater quality is to be protected or enhanced in areas where septic tank density is high.

Other Alternatives to Septic Tank Systems

On-site aerobic treatment devices are alternatives to septic tanks. These treatment devices are scaled-down versions of larger activated sludge treatment plants and employ extended aeration modes. Investigators of such systems feel that aerobic tanks, under proper conditions of design, installation and operation, release significantly higher quality effluent than septic tank systems. The main problem, however, comes from inadequate installation, maintenance and upkeep of such systems.

Providing soil enhanced treatment in areas where native soils are not suitable for the disposal of effluent from septic tanks, or where shallow soils are underlaid by till, creviced or channeled rock, or a high groundwater table, can be accomplished by use of an artificially constructed aboveground mound. Mounds, however, must also be properly constructed and maintained. Other alternatives to conventional disposal of wastewater via septic tank systems include the use of recirculating sand filter treatment systems. These systems consist of a septic tank, a recirculation tank, and an open filter. They are also economical when combined with proper disinfection (i.e., the effluent can meet surface-water discharge standards of regulatory agencies).

Also available for on-site liquid waste disposal are flow reduction and self-contained devices such as incinerating toilets, composting toilets, biological toilets, and vacuum system toilets. While these devices appear to hold many promises of adequate disposal for the future, their present use should be restricted to situations where on-site waste disposal methods are not feasible, because of natural conditions, restrictive codes, or other justifiable reasons.

Regulation of Systems and Institutional Arrangement

The life span of septic tank and leachfield systems encompasses three essential phases that can be subject to regulation (U.S. EPA, 1977). These include:

(1) Installation
(2) Operation-maintenance
(3) The ability to detect system failure and make corrections

Regulation of the installation, design, and siting, exists in most states in the U.S. Site evaluations and issuance of permits for systems are usually handled by state and local governments, although operation-maintenance of systems is largely not regulated and usually left to the discretion of the homeowner. Some states and local areas also provide promotional material on how to properly maintain and keep up systems.

Detection of failed systems and correction of them is difficult to regulate but can be somewhat handled on an individual complaint basis or when related specific health hazards exist. Protection of groundwater quality is best accomplished by regulating the installation of systems. Cesspools, where regulations exist, are generally not approved nor acceptable for any type of installation. Privies or outhouses do not constitute a significant threat to groundwater except in specific instances (i.e., close to surface waters, water supply wells) and their importance is not discussed here.

Regulation by septic tank installation as previously described is, in most cases, either by state, county, town, regional authority or a joint effort by health and/or plumbing authorities. States, however, additionally may regulate septic tank installations or only those serving commercial areas other than single family residences. States may also regulate installations constructed in critical areas. They may also delegate regulatory responsibilities, give legislative authority to local governments, or have no regulations at all. Where regulations do exist at the state level, inspections may be comprehensive or non-existent.

Some states now restrict septic tank system installation to non-subdivision situations, e.g., individually scattered lots. Mississippi, for example, does not approve individual on-site residential liquid waste disposal systems in new subdivisions, additions to existing subdivisions, or undeveloped portions of existing subdivisions, unless the establishment of a community sewage system is economically unfeasible (U.S. EPA, 1977). Louisiana also has resolved that every effort should be made to prevent use of individual sewage disposal systems in land development areas involving urban-sized lots. Individual facilities in this case are temporary and are to be replaced with proper community facilities within a short period of time (U.S. EPA, 1977). In New Mexico, lot sizes differ, depending on the type of soil involved and whether on-site or community water is provided to the property.

As states begin to evaluate groundwater contamination problems in their areas of jurisdiction which have resulted from on-site domestic waste disposal systems and practices, new regulations and controls will be developed. For example, the Oregon Department of Environmental Quality updated its on-site disposal facility regulations based on engineering studies of problems in that state (U.S. EPA, 1977), and the regulations took into account on a state-wide basis factors such as regional soil conditions and climate. Updating of their regulations can now become an ongoing process of amendments, when necessary.

The pattern of regulation of installation of septic tank/leachfield systems that has emerged in many states throughout the U.S. relegates responsibility to local health departments. However, the laws of most states still provide for enforcement intervention in cases where the activity is judged to be detrimental to the health of individuals or the community. Controls of on-site liquid waste disposal systems now move towards a clearer definition of responsibility as they revise their statutes to provide for groundwater protection, usually through amendments covering on-site sewage disposal facilities. The mandated enforcement of amended procedures is usually meant to protect potable water supplies, however, groundwater aquifers are also kept free of contamination in the long run.

Another approach for alleviating domestic on-site liquid waste related to groundwater contamination problems by regulatory agencies is the establishment and frequent use of broad-based physiographic data summaries for use in land-use planning activities. Regulating density of on-site liquid waste disposal systems based on variables such as soils, geology, physiography, hydrology, vegetation, and climate are now occurring. Technical analyses and interpretation of these data also result in an overall physical capability rating that when reduced to map form, can be used by planners and regulatory agencies.

By applying the previously described wide range planning data to problem areas, maximum densities of on-site liquid waste disposal systems could be established or amended to restrict the installation of new septic tank and leachfield systems to areas which have not yet been subjected to critical densities of residential units. As these areas then reach critical septic tank density (i.e., if further residential construction were to occur), then alternate or advanced methods of sewage disposal would be required. However, areas which have already exceeded the established critical septic disposal system density could be individually evaluated to determine what corrective measures might be taken, although by this time centralized sewerage facilities could be necessary (U.S. EPA, 1977).

FOUR

Land Disposal of Solid Wastes and Groundwater Contamination

Landfills

To evaluate effects of land disposal of solid wastes in the context of "landfills," it is necessary to recognize an unfortunate lack of distinction between properly designed and constructed sanitary landfills and the variety of operations classed as refuse dumps. A landfill is defined as "any land area dedicated or abandoned to the deposit of urban solid waste regardless of how it is operated or whether or not a subsurface excavation is actually involved" (U.S. EPA, 1973). Municipal solid waste includes household, commercial, and certain industrial wastes of which the public sector assumes responsibility for collecting and disposal. However, commercial and industrial non-hazardous solid wastes presently collected and hauled by private firms may also be discharged into public landfills, along with refuse wastes which citizens deliver themselves.

A survey conducted by the Glass Container Manufacturers Institute (*Environmental News Digest,* Nov.–Dec., 1970) reported that in 1970, 8 percent of the estimated 250 million tons of municipal solid waste generated went into properly constructed sanitary landfills, with 75 percent going into refuse dumps which were environmentally unsatisfactory. The remaining 17 percent of the solid waste material was incinerated or composted. Based on these figures, only a few citizens ever saw a landfill, hence the term generally, to them, evoked the image of a "dump." It is then easy for individuals and often public agencies to infer that leachates from landfills are ever present. However, leaching from properly designed and operated sanitary landfills does not always occur, whereas leaching from open dumps or those located in improper areas is real and extensive. Therefore, in evaluating and discussing protection of groundwater quality through control of leachate from refuse, a distinction is made between control measures built into new or existing properly engineered landfills and con-trol measures focused on dumps and poorly engineered or environmentally located landfills.

Environmental Consequences of Landfills

The hazard of landfills to groundwater quality via leachate relates to the total amount of waste generated, its spatial distribution, composition of material, siting, design, operation of the fill, and its closure and post closing plan. Because wastes are generated and disposed of usually where people are, population distribution patterns give a clue to location and intensity of landfill practices.

Typical values of components of solid wastes collected in urban communities show that slightly over 70 percent of domestic refuse can be considered biodegradable organic matter, of which about three-quarters is paper and wood (U.S. EPA, 1973). An additional fraction ranging from 1–15 percent of refuse, also involves materials which include leachate solids, such as ashes and certain soils. Studies conducted in Berkeley, California in 1952 (and repeated for the same area in 1967) verify that the percentages of individual components of refuse changed little over fifteen years of external conditions (e.g., precipitation, etc.) remain the same (Golueke and McGauhey, 1967; U.S. EPA, 1973).

Data on the amount and composition of industrial solid wastes and on their disposal, however, are less extensive. A survey of 991 chemical plants (Manufacturing Chemists Association, 1967) showed that 75 percent of solid wastes were non-combustible process solids and 71 percent of the total was disposed of in landfills on company owned property. No data are at hand on the composition of these particular wastes, but it must be presumed that some fraction would become leachate if conditions leading to the formation of it occurred. This, of course, is not the case today. Many "dumps" abound, some with high levels of toxic wastes, and we are only beginning to understand the implications of the leachate generated from these sites.

Leachate

Leachate, by definition, is contaminated water which is produced when water percolates through wastes in land disposal sites. As leachate is produced, it picks up various minerals, organics, and heavy metals. Depending on the types of wastes disposed of at the sites, the leachate developed could also contain bacteria, viruses, explosives, flammables, and various toxic chemicals. Unless these leachates are contained, the contaminants eventually migrate various distances based primarily on: magnitude of landfill operations; physical, chemical, and biological properties of the contaminants; and hydrogeological and soil conditions around the site.

If leachates migrate from the site, they contaminate ground or surface water. This then results in damages, such as contaminating private and/or public water supply wells. Contamination in these instances directly correlates with the location of the water supply source, the direction of groundwater flow, and the distance to the site.

Contaminants in a sanitary landfill are dependent on refuse content, spatial distribution of refuse based on compaction, and time variation of moisture accumulation within the landfill (Remson et al., 1968).

LEACHATE CALCULATIONS

Leachate production from a sanitary landfill penetrated by rainfall in arid or high precipitation areas can be predicted by Moisture-Routing Techniques. A Moisture-Routing Model as described by Remson et al. (1968), is based on the equation of continuity to predict leachate quantity which would be produced in a landfill for a given refuse, soil, and precipitation pattern. The model assumes that a vegetated surface with plants whose roots draw water from all parts of the soil cover exist, but that no moisture is removed by diffusing gases, that infiltration of rainfall occurs, that the soil cover and the refuse within the site are uniformly distributed, that the hydraulic characteristics are the same in all directions, and that the landfill and substrata are underlain by a groundwater aquifer.

According to the model an eight foot lift of refuse with two feet of earth cover takes from one to one-and-one-half years to reach field capacity and produce leachate if forty-four inches of rainfall are allowed to penetrate the fill material.

Leachate is not produced in a landfill until fill material reaches the water saturation point. To accomplish this, 1.62 inches of water per foot of depth of fill is necessary. This value then exceeds that which is produced from typical mixed refuse, which has about 20 percent moisture by weight (Kaiser, 1966). A moisture content of 50–60 percent induces composting, hence, a fill with little moisture (i.e., except that of urban refuse) decomposes slowly and produces low leachate volumes. However, if the initial fill material includes 80–90 percent moisture, anaerobic decomposition proceeds, causing leachate and methane production.

LEACHING OF LANDFILLS INTO GROUNDWATER AREAS

For landfill leachate to enter underlying groundwater areas depends upon several factors (U.S. EPA, 1973). These factors, together with measures for control of leachate, have been summarized by Salvato et al. (1971). Possible sources of water in landfills also include (U.S. EPA, 1973): precipitation, moisture content of refuse, surface water infiltrating the fill, percolating water entering from adjacent land areas, or groundwater coming into contact with the fill material.

Leachate is, however, not produced in landfills until a significant portion of the fill material reaches "field capacity." To reach field capacity, 1.62 inches of water per foot of fill is necessary (Qasim and Burchinall, 1970). This field capacity value is in excess of the moisture produced from typical mixed refuse.

A composite sample from average municipal refuse follows.

Composition of Municipal Refuse	Percent
Moisture	20.73
Cellulose, sugar, starch	46.63
Lipids	4.50
Protein	2.06
Other Organics	1.15
Inerts	24.93
TOTAL	100.00

To induce composting, a moisture content of 50–60 percent is required. A fill having no source of external moisture except that mixed with urban refuse decomposes slowly and produces little leachate. However, if a fill were composed of fruits, vegetables, or other high moisture products with 80–90 percent water, anaerobic decomposition would again proceed rapidly, causing leachate to be produced. Landfilling, therefore, is not recommended for large amounts of cannery wastes.

In areas of the Pacific Northwest where rainfall occurs daily during winter and in other regions of the nation where rains during summer are intense, it is difficult to place refuse in a fill without it becoming saturated with water (U.S. EPA, 1973). It seems that spreading refuse on a working face and compacting it causes the material to be exposed to moisture before it can be finally compacted and covered. Special control measures are then required for these areas to deal with leachate.

Once unsaturated refuse is deposited into a properly designed and constructed landfill, percolating water from surrounding land is not likely to enter the system (U.S. EPA, 1973). The important factor is to deposit refuse into the fill, compact it (to 750–850 pounds per cubic yard), and cover it without allowing saturation of the material to occur. Consequently, a properly constructed, compacted, and filled landfill that does not intersect groundwater will usually not cause water quality degradation. However, this is not usually the case today because of many improperly constructed systems in the past where eventually, contamination of groundwater by leachate, does occur.

If the fill is in a subsurface excavation, percolate moves downward to groundwater at a rate governed by the degree of clogging of the underlying and surrounding soil. Clogging, however, reduces permeability at the infiltrative surface, and it cannot be assumed that the landfill discharges leachate at an appreciable rate for it may also become a basin filled with only saturated refuse and soil (McGauhey and Krone, 1967). Further rainfall on the fill area will then run off the surface without coming into contact with the inner refuse. If leachate is then produced within the fill, this same soil clogging mechanism then controls its escape to groundwater. It could, however, with time, discharge a significant volume of leachate to the groundwater.

The 75 percent of urban refuse placed in dumps which are open to external sources of water is likely to produce leachate in significant amounts. It is estimated (Salvato et al., 1971) that of forty-nine annual inches of rainfall in New York, 45 percent infiltrates into unsealed and unprotected dumps, although during certain seasons of the year up to 70 percent of the infiltrated water is returned to the atmosphere by evapotranspiration. The remaining 30 percent of the infiltrate percolates through the landfill.

Not all unsatisfactory landfills were constructed in subsurface excavations. Many were built in ravines or above original land surfaces and in these landfill areas clogging beneath the fill was not the controlling factor of leachate formation. Infiltrated water in these fills tended to outcrop laterally on the surface as leachate, and flow on the surface contributing to a surface stream system (U.S. EPA, 1973).

The amount of water that passes through an open dump is significantly higher than the flow through a properly constructed sanitary landfill. Once field capacity is reached in a dump, thirty-six inches of rainfall per year that falls upon this open refuse can percolate 1 million gallons of contaminated water per acre (Salvato et al., 1971), although, in reality, this amount is reduced by surface evaporation.

Another phenomenon associated with landfills not subjected to specific control measures is the generation of carbon dioxide in the fill and its movement outward into surrounding soils. Then, when the carbon dioxide is picked up by percolating surface or rain water, the aggressiveness of this water to limestones and dolomites increases. This in turn also elevates the hardness of groundwater (U.S. EPA,

1973). Long-term environmental consequences of improper landfill practices cannot always be fully evaluated. Research is still needed before leachate as a contaminant of groundwater can be fully assessed and further understood. This becomes even more complex when one evaluates breakdown products of organic material and once deposited toxic substances.

NATURE AND AMOUNT OF LEACHATE

Data on the analyses of leachate vary widely. Much of the information comes from short-term studies in which researchers made special efforts to saturate refuse experimentally and produce maximum leaching (U.S. EPA, 1973). Experiments were then terminated before leaching rates reached equilibrium.

Studies of landfills (Emrich and Landon, 1971; California State Water Pollution Control Board, 1954; from U.S. EPA, 1973) show that leachate includes:

COD	8,000–10,000 mg/l
BOD	2,500 mg/l
Iron	600 mg/l
Chloride	250 mg/l

The California data also show that continuous flow through one acre-foot of newly deposited refuse leaches out (i.e., during the first year) approximately:

Sodium + potassium	1.5	tons
Calcium + magnesium	1.0	tons
Chloride	0.91	tons
Sulfate	0.23	tons
Bicarbonates	3.9	tons

Field studies of the amount and quality of leachate from properly designed fills were reported by the Los Angeles County Sanitation Districts (U.S. EPA, 1973). Underdrains were installed beneath two large fills at their Mission Canyon landfill to entrap leachate (Dair, 1969; Meichtry, 1971). One underdrain was installed in 1963 and the other in 1968. Initially, at the time of Meichtry's report (1971) the first of these underdrains had produced nothing but odorous gases, although the fill material had been heavily irrigated since 1968. The second, deeper, sanitary fill produced odorous gases but no leachate until March 1968 when 4.35 inches of rain fell within a 24-hour period. On that specific occasion 213 gallons of leachate were collected. Flow then continued at the rate of 1,500 gallons per month. However, periodic analyses of leachate materials indicated that a spring in a nearby canyon was contributing to the leachate volume, rather than moisture from direct infiltration of water on fill material.

Leachate materials reaching groundwater also depend on their passage through soil systems (U.S. EPA, 1973). Reduction in leachate BOD values were observed to be 95 percent during movement of liquid through 12 feet of soil (Emrich and Landon, 1971). An increase in dissolved minerals also occurred from leachate infiltrating surface soils, i.e., the aerobic soil systems stabilized organic matter (McGauhey and Krone, 1971). Fills producing leachate that is not discharged to soil surfaces develop anaerobic zones of clogging. Leachate which passes through these anaerobic zones are high in BOD, COD, tastes, and odors. These leachate criteria change little in groundwater unless it becomes diluted, withdrawn through a well, or becomes reseeded with aerobic organisms. However, if groundwater comes in contact with refuse or leachate medium, any rise and fall in the water table has a surging effect that accelerates the mixing of groundwater and leachate.

In one field observation study, a landfill partly inundated by groundwater was investigated by Hassan (1971). Well water 325 meters down gradient from the fill showed leachate to have high hardness, alkalinity, Ca, Mg, Na, K, and Cl levels, whereas at a distance of 1,000 meters these same chemical levels were undetectable. The groundwater in this study was not seriously affected. However, similar studies in Germany revealed detrimental leachate effects to groundwater areas 3,000 meters away. Leachate can also be expected to change as years pass because the present high use of garbage grinders in homes and institutions has significantly reduced the BOD concentrations of leachate (U.S. EPA, 1973).

Industrial and hazardous waste disposed of in landfill sites on property belonging to various companies has in many instances also caused many problems. This is because soluble minerals are common materials which are leached from industrial waste fills. Various toxic substances, oils, process sludges, and salt solution wastes from lagoons and pits are significant contaminants in terms of industrial waste.

A comprehensive review up to 1971 of groundwater contamination due to municipal "dumps" was prepared by Hughes et al. (1971). Topics covered included: groundwater contamination by solid wastes, research in this field at the time, regulations, criteria for site selection, safeguards, and methods of observation, detection and identification of contaminants. Two bibliographies of over 600 references complete the report.

LEACHATE PRODUCTION IN ARID AREAS

Factors affecting percolation rates through landfill sites and their relationship to leachate generation and movement show that the substance will be generated in humid areas or arid climates if landfill areas are irrigated substantially. If an average of 84–96 inches of water per year is applied via artificial irrigation or occurs from rainfall, and since evapotranspiration is never 100 percent, percolation through landfills in these instances will occur. Therefore, if irrigation systems are used to artificially water vegetative areas on the top of landfill sites, specifically in arid areas, they must be properly designed and managed for leachate collection, since it will be produced.

Routine irrigation of landfill cover areas causes leachate formation and migration. However, the major problem that will occur in arid climates during irrigation of a landfill is an increase in the decomposition rate, which then results in large amounts of methane being generated. On the opposite extreme, studies by Hughes and Cartwright (1972) concluded that leachate would not be produced in landfill areas that have low ground or surface water infiltration, i.e., particularly if the rainfall in the area is not substantial enough to infiltrate landfills, although flooding of an area with large amounts of water will cause leachate to be produced (Mertz and Stone, 1967).

Wei-chi Ying et al. (1987) describe the treatment of leachate at the Hyde Park Landfill site located in an industrial complex in the northwest corner of the town of Niagara, New York. The Hyde Park Landfill was used from 1953–75 as a disposal site for an estimated 73,000 metric tons of chemical waste, including halogenated organics. A compacted clay cover was placed over the landfill in 1978, and a tile leachate collection system was installed in 1979.

Control Methods for Leachates

Procedures that control leachate formation in landfills are those which exclude water from the area. These procedures either prevent leachate from percolating to the groundwater or collect it and subject the liquid to biological treatment (U.S. EPA, 1973). The utilization of these approaches should be maximized during the design and closure phases of landfill operations.

In existing landfill areas, groundwater contamination can be limited by incorporating the following procedures (U.S. EPA, 1973); separating wastes that are unacceptable to landfills, controlling haulers and materials they carry to the site(s), and requiring permits, registrations that specifically recognize haulers of industrial wastes and assuring that hazardous wastes are being deposited in approved sites only.

In the case of new or projected landfill sites, control measures to prevent groundwater contamination should include initially selecting sites that meet general regulation and specific objective criteria, i.e., those which recognize different classes of landfills for different kinds of waste, such as those which accept all types of hazardous waste by reason of their geological isolation from contact with groundwater, or those which accept normal mixed municipal refuse, and which accept inert earth-type materials, only.

Specific siting of landfills should also involve evaluation of alternate locations by environmental health professionals, hydrogeologists, and engineers to determine location, movement, and depth of groundwater in the vicinity; the importance of underlying groundwater as a resource (i.e., both present and future); the specific nature of the geology at the proposed site (i.e., location of fault lines, zones, etc.); and the feasibility of completely excluding surface and groundwater from reaching the finished fill. The designed landfill should also be located close to earth fill that can be used to properly seal walls and the bottom of the fill site, if the site is above groundwater. The installation of underdrainage systems that collect leachate and deliver it to sumps and drainage sump pumps to the surface, which collect, deliver, and accumulate leachate for additional treatment, should also be explored.

Sanitary landfills should be constructed with the purpose of keeping minimum refuse surface areas exposed to rainfall. Working surfaces and sites should be well drained. Dikes and fill techniques should also be used, if necessary, to isolate material from unfilled areas. Water can, however, be utilized for dust control during construction of fills. It should be used in amounts that evaporate rather than infiltrate into the ground. Surface water that accumulates during and after construction of the fill should also be diverted. This can be accomplished by use of peripheral bypass drains. The fill should also be compacted at a proper angle and slope, so adequate surface drainage occurs. Venting of gases through the fill cover should also be engineered properly. In some situations it may also be necessary to bale refuse and have a fill composed of compacted bales which can then promptly be covered and their surface drained.

In "new or existing landfills" the U.S. EPA (1973) recommends that provisions be made to continue periodic maintenance of fill cover. This can be done by regrading the surface, since shrinkage of refuse eventually causes cracks, and depressions on cover areas will increase infiltration. If the cover is to be seeded by vegetation, it should be done with use of high transpiration crop seeds. Excessive irrigation of surface plantings in this case should not be allowed. In addition, surface and groundwater volumes should be diverted around fill sites and not through them. The amount of putrescible solid waste could also be reduced by initiating or participating in regional reclamation activities that feature energy conversion of reusable refuse.

In "the case of existing landfills and dumps" U.S. EPA (1973) recommended that if contaminated groundwater does occur next to fills, that it be intercepted at the fill site by constructing wells at or near fill areas. Water obtained from these wells can then be cleaned and returned back to the aquifer.

Statewide programs of waste management which feature use of regional landfills in adequate geological areas, should also be initiated and implemented. Regional landfills, when constructed, will replace numerous small refuse dumps with large approved landfill systems and eventually cause phasing out of any leachate contribution to groundwater that these small dumps in time would have had.

Of the previously described potential control measures, those which are applicable to new, approved sanitary landfills will prevent or eliminate groundwater contamination by leachate. That is why adequate siting, construction, operating, and proper maintenence of fills are also considered control measures. Properly engineered landfills, although not usually equipped with underdrains, will have minimal detrimental effects upon groundwater quality. Similarly, improperly constructed or engineered landfill areas have already contributed large amounts of contaminated leachate to groundwater aquifers.

Recognized hazardous waste streams are now treated under a different system of landfilling. It is not within the scope of this book to discuss substantially hazardous waste streams, although the subject is presented to some degree in chapter 13. Farrara et al. (1984) present a comprehensive review of the disposal of hazardous wastes and their relationship to groundwater quality.

FIVE

Municipal, Industrial, Oil Field Wastes, and Their Effects on Groundwater

Municipal Sewage Disposal Systems

The potential of municipal sewage disposal systems to contaminate groundwater varies. Municipal sewers are intended to be watertight and present no hazard to groundwater, except if temporarily disruption occurs accidently. In reality, however, leakage does occur, especially from older systems. Leakage in gravity sewers can result from the following listed causes (U.S. EPA, 1973):

- poor workmanship, especially at the time cement mortar was applied to join material
- cracked or defective pipe sections incorporated when constructing sewer lines
- breakage of pipes and joining material by roots from trees that penetrate or move sewer lines
- displacement and rupturing of pipelines by superimposed loads, heavy equipment, or earthfill on top of pipe laid in improperly constructed foundations
- rupturing of pipe joints and pipe sections by slippage of soil in hilly topographical areas
- fracturing and displacement of pipe by seismic activity (e.g., earthquakes)
- loss of foundation support because of underground washout
- shearing of pipes at manhole cover areas due to differential settlements
- sewage backup moving into abandoned sewer laterals and causing breakages in lines

Environmental Consequences

Effluent that escapes from sewer lines via leakage routes is raw sewage. This sewage may be actively decomposing and mixed with industrial waste chemicals. If a sewer line, therefore, is deep underground and close to groundwater, contaminants will be released below the biologically active zone of the soil and introduced into the receiving groundwater. Chlorides, bacteria, viruses, and unstable organics, trace metals, and other chemicals that produce foul tastes and odors are then introduced this way.

If a fractured sewer line is below the groundwater table, infiltration of water into the system, rather than leakage, will result. However, if infiltration occurs only during parts of the year, the intermittent flows may unclog the system and consequently a higher seasonal leakage rate will occur than what would have been the case on a year-round basis.

Outfall sewers located under high pressure areas are constructed of cast iron, steel, transite, concrete pipe, or other strong materials. These outfall sewer systems have fewer joints per mile than gravity sewers and the joining is less likely to be of poor quality that readily ruptures. Because of their superior construction and engineering attention, outfall sewers are not a threat to groundwater quality. However, when leakage does occur, it is of an outward nature regardless of whether pipes are above or in groundwater. Because of high pressures, small openings between sections eventually clog, but before then, sewage is usually injected into soil or groundwater via direct means. Effluent from these high pressure systems then appears on the surface of the soil or shows up as outcrop on hillsides. It is in these outcrop areas where it can easily be detected by sight and odor. Storm drains, on the other extreme, slow the flow of surface water to surface streams, because there is no direct infiltration. Storm drains, however, have the potential to spread oils and soluble matter from streets, fertilized land, and pesticide-treated gardens and lawns onto areas that feed into surface water or holding basins.

Reasons, however, why leakage from sewer lines will continue to decrease include the fact that (U.S. EPA, 1973):

- New sewers and replaced old lines are now laid with joining materials which are watertight and can

37

adjust to changes in pipe alignment without fracturing.

- Construction methods have improved and a better product is produced by contractors.
- More rigid specifications and better inspection procedures provide a finer end product.
- Equipment for photographing or televising interior areas of pipelines is available and used to locate leaks and fractures in a sewer system, thus preventing bigger problems later on.
- Public works departments from various municipalities use their own personnel or private contractors to systematically survey the condition of sewers within their jurisdictions (i.e., a "search and destroy" program of sewer repair and maintenance occurs).
- Sewer maintenance by itself is a special division of many public works departments, and annual short courses and training programs in the technology of maintaining sewers are available nationwide.

Control Methods

Procedures for controlling sewer leakage to groundwaters are implicit in improved practices as previously listed, although the sewerage system of large communities can be an infinite point system of possible leaks. The system, however, is underground and not subject to surface observation and repair for control purposes. Therefore, a productive control program should include more than the correcting of problem areas when they are "uncovered." It should include the following features (U.S. EPA, 1973):

- a publicized policy of maximum protection of groundwater resources as part of an overall concern for natural resource conservation
- organized, identified, and responsible staff resources for sewer construction and maintenance
- modern codes and specifications for sewer construction, including maintenance and appropriate systematic inspection procedures
- systematic internal and external inspection of existing sewer systems at five-year or less intervals to detect, repair leaks, or replace sections of sewer lines
- continual training of sewer maintenance personnel
- exclusion from discharge to municipal sewers of hazardous materials or toxic substances

As with lagoons, basins, and pits, monitoring of groundwater quality in relation to sewer leakage can be accomplished by a program of collection and evaluation of relevant data specific for each metropolitan area. Continued surveillance and control procedures that work should be part of a routine prevention and correction leakage program.

Municipal Wastewater Effluent

Effluent from municipal wastewater treatment plants often is discharged into surface waters after some type of treatment and their meeting of regulatory standards (NPDES). However, in some instances the treated water is reclaimed by allowing it to percolate into the ground to recharge the aquifer or is applied to soil surfaces. Usually, other regulatory standards also have to be met at this time.

Studies Concerned with Land Application of Wastes and Recharge

Groundwater contamination possibilities inherent in artificial water recharge methods via use of municipal discharge have been reported by: U.S. EPA (1974), Popkin and Bendixen (1968), Tchobanoglous and Eliassen (1969), Bouwer (1969), Born and Stephenson (1969), and Bernhart (1973). In addition, the University of California Sanitary Engineering Research Laboratory gathered and evaluated pertinent studies, methods, and statistics of water recharge into aquifers by effluent spreading and injection (U.S. EPA, 1974).

Popkin and Bendixen (1968) summarized studies concerned with the application of municipal liquid waste to soil. Continuing hydraulic acceptance and percolate qualities were stressed. Design and operation of soil adsorption systems could be improved by weekly dosing and/or use of improved pre-treatment processes.

Tchobanoglous and Eliassen (1969) described various factors related to the indirect cycle of water reuse. Methods of treated municipal wastewater recharge were discussed. They included surface spreading, direct injection, and use of pits or leachfields for effluent seepage. A cost-benefit analysis for economic feasibility of such an indirect reuse of reclaimed water was also described by Tchobanoglous and Eliassen (1969).

Bouwer (1969; from U.S. EPA, 1974) emphasized how aerobic percolation and subsequent lateral movement of low quality water removed biodegradable materials, pathogenic organisms, and certain inorganic substances. However, nitrate reduction by denitrification microorganisms was identified as a major problem in renovating sewage effluent.

Liquid waste disposal by irrigation and subsequent geohydrological concerns have also been reviewed by Vorn and Stephenson (1969; from U.S. EPA, 1974). The thickness, nature, and distribution of unconsolidated surface deposits determine infiltration, adsorption storage, and downward movement of wastewater. Infiltrometer testing, laboratory examinations, and flow systems were discussed as methods of monitoring wastewater recharge.

Additional monitoring and control methods were also analyzed by Martin (1969; from U.S. EPA, 1974), with soil surveys and data used to minimize leaching, erosion, and groundwater contamination. Nitrogen contamination problems, in this instance, could be controlled by anaerobic conditions, plant growth, holding lagoons, or rotation spreading. Several other successful land waste disposal systems were also described in this report.

Dvoracek and Wheaton (1970; from U.S. EPA, 1974) described how groundwater becomes contaminated by artificial recharge due to the quality of the initial recharge water. Various methods of recharge were discussed (e.g., wells, shafts, holes, pits, trenches) including land surface spreading of water. The groundwater contamination potential of each of the previously mentioned recharge methods was also evaluated.

Planning and design criteria for specific waste disposal methods for municipal areas were presented in 1972–73 by the U.S. EPA (U.S. EPA, 1974). The Pennsylvania Bureau of Water Quality Management also published a manual on spray irrigation methods and designs. Design and engineering aspects of proposed operations, site selection criteria, and essential groundwater quality monitoring data were also detailed.

Similarly, Bernhart (1973) analyzed soil infiltration and evapotranspiration methods for wastewater disposal. Design elements and area calculations for seepage beds were included. The infiltration effectiveness of septic tanks, aeration tanks, conventional tile fields, and seepage beds were charted. The study also evaluated "horizontal protective distances" required between water supply wells and sources of contamination under different conditions.

Additional localized studies of effluent recharge and disposal problems and practices have also been conducted in many states since 1950. The Pennsylvania State University studies 1967–68 dealt with renovation of wastewater effluent by irrigation of forest land (Pennypacker et al., 1967; Sopper, 1968). Pennypacker et al. (1967) conducted a field study of treated sewage effluent that had been sprayed on forest land and observed various phosphorus levels on top of the soil layers. He also showed that greater depths of infiltrations were required to remove nitrates, potassium, calcium, magnesium, and sodium from soils. Sopper (1968) in his study concerning treated municipal wastewater in forested areas achieved satisfactory renovation at rates of up to 4 inches per week. Approximately 90 percent of applied wastewater was recharged to groundwater reservoirs at an application rate of 2 inches per week.

A review of a municipal sewage disposal system for St. Charles, Charles County, Maryland, showed that wastewater in this instance had been renovated to drinking water by use of sewage lagoon systems supplemented by spray irrigation procedures (Anonymous, 1973). Bendixen et al. (1968) reported results of monitoring municipal ridge and furrow liquid waste disposal systems in Westby, Wisconsin. Four one-acre basins accommodated trickling filter effluent into the soil and a heavy stand of unharvested grass also contributed to successful operation of this particular system. Changes in infiltration rates and groundwater quality due to seasonal and different loading and operating conditions were also evaluated. Ketelle (1971; from U.S. EPA, 1974)

discussed hydrologic and geologic factors relating to land disposal of wastes. A case study in southeastern Wisconsin was presented. Geography, climate, geology, soils, and groundwater of a seven-county region were analyzed, and a final map of the area was developed. The map indicated the location and suitability of specified areas for the disposal of liquid waste effluent.

Muskegon County, Michigan, was the site of a research project undertaken by Chaiken et al. (1973) to study land disposal of treated sewage by spraying. A detailed observation well network system was established that consisted of 300 wells in close proximity to storage lagoon ditches. Groundwater samples in this study were analyzed for forty-three physical, biological, and chemical water quality parameters.

Harvey and Skelton (1968) summarized a field study at a Springfield, Missouri, sewage plant, including secondary treated effluent travelling underground where anaerobic conditions prevailed. Seepage runs, dye studies, and seismograph surveys were used to determine sources of pollution.

Brown and Signor (1972) surveyed methods of groundwater recharge in the Southern High Plains of New Mexico and Texas. Artificial recharge contamination hazards included recharge waters having high particulate matter concentrations and faulty design and construction of wells.

An "overland-flow" sprinkler irrigation system that disposed of wastewaters from a cannery in Paris, Texas, reported a 99 percent reduction in BOD (Anonymous, 1973). Up to 90 percent reduction in nitrogen from vegetable and grease wastes also occurred. The 640 acres in this study were designed to accept a wastewater application rate of one-quarter–one-half inch per day. Underground migration of contaminants did not occur due to downslope percolation, terrace collections, and channels to receiving streams.

An experimental project on reclaiming water from secondary sewage effluent with infiltration basins in dry Salt River beds near Phoenix, Arizona, was reported by Bouwer (1968; 1970) and Bouwer et al. (1972). The hydrogeology of the Salt River bed in this instance was suitable for high rate wastewater reclamation by groundwater recharge. The project contained six recharge basin areas, with infiltration rates being different in grass, gravel, or bare soil basins. Removal of 90 percent of nitrogen was obtained during long inundation periods and reductions in BOD, coliforms, and phosphorus were also observed. The studies in this locality showed that surface-spreading renovation of wastewater for groundwater recharge is comparable to tertiary in plant treatment costs.

Wilson et al. (1968; from U.S. EPA, 1974) examined the dilution of municipal waste effluents with river water within thick alluvium during pit recharges near the Santa Cruz River in Arizona. Water content profiles and groundwater hydrographs were developed for recharge sites and abutting ephemeral streams. The report concluded that recharging of highly concentrated waters should coincide with periods of fully developed river recharge times and/or when ephemeral streams were discharging. The study also recommended that additional evaluation of pumping, as a mixing and blending procedure should also be conducted. Groundwater quality monitoring and its significance was also emphasized in this study.

Artificial recharge into groundwater areas in California have been extensively documented since 1950. Stone and Garber (1954) in their study described infiltration of sewage effluent through a spreading basin in Los Angeles County. Data compiled by the authors showed that aerobic conditions in percolating fluids, successively allowed wastewater to be reused 2–5 times. Standards for dissolved oxygen and BOD contents in percolated fluids at that time were also discussed.

The California State Water Pollution Control Board reported on bacterial and chemical pollution at a wastewater reclamation project at Lodi (U.S. EPA, 1974). Bacteriologically safe water in this instance from settled sewage or final effluent resulted from passage through four feet of soil. Water of satisfactory chemical quality was also achieved if initially raw sewage did not contain high concentrations of industrial wastes. Safe percolation rates, spreading techniques, and side effects, e.g., rearing of mosquitoes, growth of algae, and liberation of obnoxious odors, during wastewater reclamation were also presented.

Industrial Contamination of Groundwater

Underground disposal problems of industrial wastes and groundwater contamination were studied by Ives and Eddy (1968; from U.S. EPA, 1974). The task group study assessed the effects of industrial waste disposal on groundwater and noted increasing threats to this resource across the nation. Statutory control measures were then recommended. Ives and Eddy (1968) in their study surveyed underground waste disposal policies and practices for various states, specifically for contamination problems of groundwater associated with subsurface waste disposal wells. Ulrich (1955) in another study described a field incident involving high chloride levels in wells at Massillon, Ohio. Chloride levels increased from 8 mg/l to 1700 mg/l with the cause believed to relate to industrial wastes infiltrating from an adjacent river. In a 1954–55 study at Indian Hill, Ohio, Parks (1959) reported brine being discharged from a water softening plant. This brine percolated 850 feet to pumping wells. The chloride concentration in the well waters rose from 20 mg/l to 744 mg/l.

Disposal practices and contamination of groundwater in the Rocky Mountain Arsenal of Colorado have constantly been reviewed since 1950. In a 1954–56 study, Petri (1962) reported that chloride levels increased in groundwater 200 percent due to wastewater infiltration from disposal ponds. The control method employed then included lining of the ponds to stop the infiltration. Walker (1961) in another study, described improper waste disposal practices as being

responsible for groundwater contamination because unintended recharge was occurring. The contaminated groundwater from artificial recharge, in this case, was toxic to crops and unsafe for human consumption.

Swenson (1962) in his study, reviewed the "Montebello incident" and its aftermath. This particular incident included a chemical plant which manufactured weedkiller in Montebello, California that discharged dichlorophenol into groundwater areas. Within seventeen days city wells were contaminated and although the plant ceased discharging wastes within thirty days, foul tastes and odors persisted for five years in well waters.

Evans (1965) presented an analysis of the dangers of industrial waste treatment and disposal facilities. Emphasis was placed on the diverse nature of industrial wastes and on the effects of these wastes on groundwater when disposed of into deep wells and lagoon systems.

A study of the groundwater quality in the Fresno-Clovis area conducted by the California Department of Water Resources showed that groundwater near the Fresno sewage treatment plant, where effluents were discharged under the land surface, was of lower quality (U.S. EPA, 1974). It was concluded that this contaminated groundwater slug would move toward the city as long as the water table continued to be lowered.

Price (1967) reviewed groundwater contamination problems in an alluvial aquifer in Keizer, Oregon. Industrial waste from an experimental aluminum reduction plant had been dumped into a borrow pit, and sulfate concentrations exceeded 1,000 mg/l. Water samples from wells were analyzed for hardness as the principal indicator of contamination. Although diluted in the immediate vicinity of the pit, the contaminant seemed to spread within the aquifer for more than one mile.

Bergstrom (1969) described waste disposal operations in Illinois that included: landfills and dumps, radioactive waste burial sites, sewage treatment and waste storage ponds, disposal wells, and sewage-stormwater tunnels. Waste management proposals intended to protect groundwater quality in this area focused on site selection criteria, hydrogeologic data, and investigations concerning water movement in different geological strata.

Maehler and Greenberg (1962) presented data evaluating organic contaminants in groundwater. The disposal methods of petroleum industrial wastes were analyzed. The results indicated that organic analyses in contamination studies were essential in any groundwater contamination evaluation process. The wells they sampled were contaminated with organic compounds, and the compounds analyzed were direct by-products of oil field wastes.

Deutsch (1962) reported on the occurrence of phenols in groundwater at Alma, Michigan. Refinery wastes were being discharged into pits that were percolating into the aquifer. Glacial drift deposits in this area allowed vertical and horizontal migration of contaminants to occur, and two wells had to be abandoned as a result of contamination. To remedy the problem, pits were sealed and pumping from the wells was restricted to decrease the hydraulic gradients. Phenol, however, continued to be present in small quantities.

A detailed statistical summarization of worldwide groundwater contamination due to petroleum products has been reported by Zimmerman (1964; from U.S. EPA, 1974). Sources, extent, and effects of petroleum products causing contamination are detailed. Control and detection techniques are also included.

Grubb (1970) described a break in an industrial waste discharge line, which when coupled with a forty-nine-foot rise in the Ohio River, allowed hydrochloric acid to enter an outwash aquifer used by a Kentucky industry. Chloride concentrations in excess of 30,000 mg/l were observed in water discharged from the industrial well nearest the break, and within a year that well had to be abandoned. Additional fluctuations of chloride levels in an industrial well near the river indicated that highly mineralized groundwater occurred near the acid source.

Van der Warden et al. (1971) reported on laboratory studies conducted on the transfer of hydrocarbons from residual oil zones to trickling water areas. The study modeled oil spills in soils including the transfer process of oil components to groundwater. The transfer of oil components was delayed by adsorption onto soil and their concentration in drain water also decreased. Adsorption effects under field conditions appeared strong, and oxidation and evaporation also seemed to determine the fate of oil in soil.

Williams and Wilder (1971) reviewed the effects of a gasoline pipeline leak near Los Angeles, California. A large amount of gasoline (i.e., 250,000 gallons) seeped into a groundwater supply system. Extensive analytical studies of the fluid systems, including clean up and restoration attempts were enacted. By use of skimming wells 50,000 gallons of free gasoline were removed from the aquifer.

A survey of petroleum contamination of groundwater areas in Maryland was presented by Matis (1971). Most counties in Maryland recorded contamination. Problems, however, were usually localized because it is difficult to remove petroleum products from groundwater. Legal and regulatory problems throughout the state also continued long after the original complaints ceased.

Collins (1971) in his studies, undertook a review of contamination potential of oil and gas during well drilling and development. Mechanisms by which brines, crude oil, or gas infiltrate and contaminate groundwaters were described.

The Committee on Environmental Affairs of the American Petroleum Institute published a survey on the migration of petroleum products in soil and groundwater (API, 1972). Petroleum product contamination was discussed and several incidents of oil pollution were described. The report also contains material on controlling oil spills, recovery of oil, and detection of hydrocarbons.

Metal Wastes

The discharge of plating wastes by aircraft manufacturing plants in Nassau County, Long Island, New York, was reported by Davids and Lieber (1951). They examined the problem of groundwater contamination by chromium from release of materials into diffusion wells and shallow pits. After this incident, large industrial consumers of chromic acid were required to install treatment facilities to remove hexavalent chromium from their waste streams prior to disposal. The results were that chromium was almost completely removed before it was allowed into the ground.

Lieber and Welsh (1954), in a similar study, reported on cadmium levels of 0.01–3.2 mg/l in groundwater of the previously described area. The apparent path of the cadmium contaminant also coincided with the direction of groundwater flow. The appearance of cadmium as a groundwater contaminant during this time created a need to ascertain safe and reasonable limits for the element in potable water systems. Lieber et al. (1964) also conducted a field study of both cadmium and hexavalent chromium in groundwater areas near South Farmingdale, Long Island, New York. Groundwater contamination from a metal plating plant was summarized and the status of contaminant products was evaluated through the use of ninety observation wells with water samples being taken at five-foot intervals. Improved treatment facilities at the source of the cadmium and hexavalent chromium contamination turned out to be the best method of control.

A similar study in the same area by Perlmutter et al. (1963) exemplified the movement of a slug of contaminated groundwater beneath a metal plating plant. Concentrations of chromium and cadmium were high. The identified slug then became diluted as it discharged into a creek. Improved treatment procedures at the plant reduced cadmium and chromium concentrations. Perlmutter and Lieber (1970) also reviewed contamination of groundwater by plating wastes in the same area and described the extent of contamination in the Upper Glacial aquifer. The seepage of plating wastes containing cadmium and hexavalent chromium into the aquifer formed a plume of contaminated water 4300 feet long, 100 feet wide, and 70 feet thick. Their report also discussed factors contributing to the longitudinal, lateral, and vertical spread of these particular contaminants.

Mine Wastes

Acid mine drainage and its associated problems in the Appalachian mountain region have been described by Rice and Co. (1969; from U.S. EPA, 1974). The effectiveness of control techniques was studied from actual contamination problems. Techniques which controlled groundwater contamination from mine wastes included: neutralization, reverse osmosis, streamflow regulation, deep-well disposal, land reclamation, re-vegetation, pumping and drainage, water diversion, mine sealing, proper refuse treatment, and impoundment of acid water.

Emrich and Merritt (1969) reported on degrading effects of mine drainage on groundwater areas in Appalachia. Oxidation and leaching chemical reactions connected with coal mining produced high iron and sulfate concentrations and low pH values in groundwater. Oil and gas wells abutting natural joints and fractures of the rocks permitted acid mine drainage liquids to move downward from strip mines into underlying aquifers that increased the iron and sulfate content of water.

Ahmad (1970) presented a plan to solve the acid mine drainage contamination problem of Lake Hope, Ohio. Coal mines in the area were continually being flushed by the disturbed natural groundwater flow system, and sulfuric acid was produced in the process. Three separate aquifers existed near the Todd mine and the plan was to discharge the uncontaminated water from the upper aquifer to the lower one. This was done to stop the flow of water through the mine. The plan worked in this instance.

A study by Mink et al. (1972) revealed high zinc, lead, and cadmium concentrations of the Coeur D' Alene District near Wallace, Idaho. Groundwater contamination in this instance resulted from leaching of old mine tailings in the upper part of the sand and gravel aquifers. This particular problem was complex because a nearby settling pond recharged the groundwater, raised the water table, and caused more leaching by influencing the vertical movement of water. Galbraith et al. (1972) also presented the results of a field study at Cataldo Mission Flats, near Coeur D' Alene, Idaho. The leaching of heavy metals by groundwater passing through mine tailings was caused by the oxidation of sulfides and the action of microorganisms. The report analyzed the chemical processes involved and the relationship between the concentrations of elements and pH.

Merkel (1972) conducted a resistivity survey in an area with good geologic control to determine if resistivity was a viable technique for delineating aquifers contaminated with acid mine drainage. Based on the results, he presented techniques for the periodic monitoring of groundwater areas using surface resistivity to determine the extent of acid mine drainage contamination.

Pits and Lagoons

Harmeson and Vogel (1963) surveyed physical, chemical, bacterial, and radioactive contaminants moving into groundwater via artificial recharge. They also reported on treating river water with chlorine and discharging it into recharge pits. Although existing water quality standards were met at the time, the presence of radioactivity in groundwater continued to pose an increasing threat. The report also contained charts concerned with a full range of water quality analyses.

Preul (1968) described field observations over a three-year period concerned with nutrient concentration in groundwaters near ten waste stabilization lagoons in Minnesota which treated domestic wastewater from small municipalities. Subsurface soils were sandy, silty, and the ponds had high percolation rates. Ammonia nitrogen concentrations were adsorbed within 200 feet of one pond. Phosphates were not observed in significant concentrations and were being adsorbed by soils over a wide pH range. Alkyl benzene sulfonate (ABS), however, was observed at levels of 0.5 mg/l, 200 feet from another pond. This was also true of other organic products resistant to biodegradation.

Hackbarth (1971) presented a method for supplementing data from piezometers to monitor waste disposal sites. Examination of a time sequence of resistivity measurements at fixed points in a disposal area was involved. Sulfite liquor movement away from a seepage pit was also studied using this method. Hackbarth's study listed conditions necessary to obtain results from piezometers.

Radioactive Materials

DeLaguna and Blomeke (1957) described disposal of radioactive wastes from processing of solid fuel and blanket elements. Their report discussed several methods of uranium extraction, including: removal of element jackets, treatment of uranium zirconium fuel elements, deep-well disposal problems, well hydraulics, thermal considerations of disposal aquifers, regional hydrology, deep-well disposal areas in the United States, and associated disposal costs.

Roedder (1959) detailed problems involved with the disposal of acid aluminum nitrate high-level radioactive waste solutions by injection into deep, brine saturated aquifers (e.g., salaquifers). The concept of a "zone of equilibration" was developed to aid in discussing mechanics of interaction of moving wastes with salaquifer minerals. The width of the "zone" controlled the usable storage capacity of the salaquifer. It was also shown experimentally that reactions occurred with carbonates, limonite, clays, and other typical salaquifer materials. These reactions caused the precipitation of aluminum and ferric hydroxide gels, and effectively blocked further injection. Only under certain very special conditions would these injection procedures be economically feasible.

Belter (1963) reported on radioactive waste management activities of the Atomic Energy Commission. Two basic approaches to effluent control were defined at that time: "dilute and disperse," and "concentrate and contain." Specific environmental aspects in waste disposal practices and the distinction between basic radiation protection standards and performance criteria of control operations were discussed. Radioactive waste disposal practices and methods for various types of wastes were also described. The status of research in the area at that time was also noted. General economic factors relating to radioactive waste handling and disposal were also included.

Champlin and Eichholz (1968) reported on the movement of radioactive sodium. Radioactive solutions were injected into a model aquifer and the appearance of radioactivity in the effluent correlated with increases in suspended particulate matter, potassium and calcium concentration, and conductivity. Their studies concluded that significant amounts of radioactivity were transported through the test bed, despite the high solubility of the sodium ion.

Witkowski and Manneschmidt (1962) described a decision that had been made to discontinue the ground disposal of liquid waste at Oak Ridge National Laboratory, Tennessee. Serious long range ground and surface water contamination problems were cited as factors influencing this particular decision.

The radioactive waste disposal facilities of the Savannah River Plant near Aiken, South Carolina, were described by Reichert (1962) and Reichert and Fenimore (1964). The hydrogeologic, climatic, and demographic characteristics of the area did not support the disposal of radioactive wastes in this area. Ground disposal, therefore, had been limited to burial of solid wastes and discharge of low level liquid wastes to seepage basins. Radionuclides had not been detected in groundwater, but strontium-90 was observed in sand layers 500 feet from the seepage basins. Radionuclide migration was slow wherever soils did not contain sandy strata or sand-filled clastic dikes.

Proctor and Marine (1965) conducted an investigation on the technical feasibility and high degree of safety attainable by storage of high level radioactive wastes in unlined vaults excavated in crystalline rock 1,500 feet beneath the surface of the Savannah River Plant. The most significant force aiding radionuclide migration from the storage site was derived from natural groundwater movement, coupled with effects due to dispersion and ion exchange. Three factors, however, prevented radionuclide migration. They included: the low permeability of crystalline rock, the impermeable clay layer separating rock and overlying sediments containing prolific groundwater zones, and ion exchange properties of the sediments. It appeared that any one of these barriers could contain radionuclides longer than 600 years, the time required to render them innocuous.

Lynn and Arlin (1962) described a deep-well injection system near Grants, New Mexico, for disposal of uranium mill tailing water. The reservoir sandstones contained water similar to injected wastewater and were isolated from overlying freshwater aquifers by an evaporite barrier zone. The mill tailing water was decanted, filtered, and introduced into the well by gravity at a rate of 400 gpm. The life expectancy of the reservoir was estimated to be ten years.

A Los Alamos, New Mexico, disposal area for liquid radioactive wastes was studied by Purtymun et al. (1966). The fine particles in the alluvial materials had a greater affinity for radionuclides than coarse particles. The radioactivity in the alluvium was dispersed by wastewater, storm runoff, and decreased with distance from the point of effluent outfall. Radionuclides were retained in the upper three feet of the deposits. Very little groundwater quality change was noted.

Since 1961, numerous studies have been conducted concerning the disposal of liquid radioactive wastes to a leaching pond at the National Reactor Testing Station in Idaho (Jones and Shuter, 1962; Morris et al., 1964; Nebeker and Lakey, 1972). Jones and Shuter (1962) described seepage from a pond in an extensive sedimentary bed 150 feet underground. The observed tritium content of the perched water was thought to be low, considering annual discharge rates. Morris et al. (1964) investigated chemical and radiometric changes that occurred as groundwater containing radioactive wastes moved through basalt and unconsolidated sediments of the area. Water level and test hole sample analyses were detailed in this report.

Geological and hydrological aspects of the disposal of liquid radioactive wastes at the then known Atomic Energy Commission's Chemical Separations Plants at Hanford, Washington, in 1956, have been the general subject of many studies since that time (Brown et al., 1956). Brown et al. (1956) reported on rate and direction of flow of groundwater in the area and effects of disposal operations. Emphasis was placed on microgeologic and microhydrologic procedures, including predictability concepts. Raymond and Bierschenk (1957) also described hydrologic, geologic, and radiologic monitoring data obtained from several hundred wells in the same area throughout a twelve-year period.

Although the U.S. EPA has until June 1989 to regulate radionuclides under the amended Safe Drinking Water Act (SDWA), the drinking water industry is already trying to eliminate contaminants.

Radionuclides are by definition nuclides that exhibit radioactivity (Aieta et al., 1987). Radioactive decay is the spontaneous transformation of a radionuclide into one or more stable nuclides or radionuclides. Radon is a water soluble gas produced by the decay of radium isotopes. The higher levels of Rn, which is the predominant form of radon, is typically found in groundwater. For example, it is estimated that as many as 10–50 percent of the 45,000 public drinking water suppliers that currently rely on groundwater in the United States will have to remove radon to be in compliance with the SDWA regulations (Aieta et al., 1987). It is recommended that water utilities with supplies susceptible to radionuclide contamination initiate monitoring programs.

Several processes have been shown to be effective for the removal of radium. Radon can be removed by granular activated carbon absorption and aeration. Additional information concerning radionuclides, including uranium, can be obtained by reviewing papers by Cothern (1987), Prichard

(1987), Lowry et al., (1987), Valentine (1987), and Wren et al. (1987).

Municipal Sludge on Land

The disposal of sludge on land from municipal wastewater disposal systems could be a means of using a plant nutrient or soil conditioner as a potential resource rather than a liability. Municipal sludge can be applied to land and measured in terms of an economic benefit. This application of sludge on land of course cannot include hazardous or toxic substances, heavy metals, and/or infectious materials because of the potential health risks involved.

It is not within the scope of this manuscript to discuss the management and disposal of municipal sludge, other than to recognize that certain toxic industrial sludges could find their way into groundwater.

Oil Field Brines

In 1967 a subcommittee of the Interstate Compact Commission published a survey of water problems associated with oil production in the United States (U.S. EPA, 1974). The nature and extent of the problems were outlined, and reports on individual state regulations and enforcement policies were described. Also during 1967, an article appeared in *Petroleum Equipment and Services* (1967) dealing with federal, state, and local regulatory agency attempts to control contamination resulting from oil field brines. The disposal techniques discussed in this report included the use of disposal wells, open pits, and lined reservoirs for control purposes. Many disposal techniques in the field of wastewater disposal were reviewed in this article.

The occurrence of groundwater contamination from natural gas and oil production activities in New York was summarized by Crain in 1969 (U.S. EPA, 1974). Leakage of natural gas wells resulted in groundwater contamination as saltwater or gas, but effects were localized and difficult to separate from natural contamination sources. The presence of oil and saltwater in groundwater was due to secondary recovery of oil by the water flooding method. Contamination from active oil fields was caused by spillage and separator units that disposed wastes on the ground. Contamination in abandoned fields was caused by upward movement of contaminants which had been under artesian pressure. This leakage occurred through uncapped or leaking wells and was expected to increase substantially.

The lack of groundwater contamination from oil wells, gas wells, and dry holes drilled in Michigan was summarized by Eddy (1965). Special emphasis was placed on plugging of wells as an important phase of groundwater protection.

Fryberger (1972) conducted a field investigation of a brine-polluted aquifer in Miller County, Southwest Arkansas. The study focused on oil field disposal pits and disposal wells with contamination expected to last 250 years. Attempts at rehabilitation of these areas included pumping the brine into the Red River and deep wells, however, these methods were not economically justified.

Knowles (1965) discussed the hydrologic aspects of oil field brine disposal in Alabama where groundwater contamination was frequent. Disposal of brines and other wastes in Alabama, specifically, the Pollard field, occurred by injection into disposal pits and evaporative pits. Proper maintenance and lining of pits to prevent brine and oil leaks were effective methods of preventing contamination.

Irwin and Morton (1969) produced hydrogeologic information on the Glorieta Sandstone and Ogallala formations as related to underground waste disposal. Permits for 147 oil field brine disposal wells had been issued in the area, and increased vertical permeability between the formations resulted in upward movement of brine under hydrostatic head from the Glorietta sandstone formation into the overlying freshwater aquifer, the Ogallala.

McMillion (1965) surveyed hydrologic aspects of disposal of oil field brines in Texas. Some brines were disposed of into unlined earthen pits that seeped, overflowed, and contaminated freshwater. Others were injected into subsurface areas where inadequate well completion methods constituted a longer-term problem than surface disposal. The need for brine contamination control programs of groundwater with the objective of enhancing maximum oil and gas conservation and development was emphasized.

Payne (1966) described brine injection and its effects in Texas. Drilling, production, and abandonment problems concerned with oil and gas operations were discussed. Guidelines were presented for design, operation, and control of effective saltwater injection systems.

Burke (1966) reviewed the Texas Railroad Commission regulations of 1965, which imposed controls over saltwater disposal. The Commission's regulations placed greater emphasis on proper well completions, encouraged controlled subsurface injection, eliminated all earthen pits, enforced plugging procedures, and closely inspected completion techniques of water injection wells.

Rice (1968) described saltwater disposal techniques in the Permian basin of Texas and New Mexico. Systems of reinjecting produced water into brine-bearing formations were summarized. Various design, installation, and operation factors were considered for safe and economical systems.

McMillion (1971) conducted a field investigation of mineralized groundwater from oil field brine pits in eastern New Mexico. Poor quality groundwater was pumped and used for the secondary recovery of oil. Three corrective measures for the thousands of discontinued brine pits in the region were offered. These measures included: restriction of freshwater pumping in zones where groundwater movement would influence contaminants, removal of highly mineralized water from the area, and developing and pumping poor quality groundwater to hold its movement in check.

Rold (1971) surveyed groundwater contamination problems in the "oil patch" of the arid West. The sources of groundwater identified by Rold included evaporation pits, insufficient surface casings, injection systems, abandoned wells, and seismic shotholes. Constant planning, monitoring and policing of oil field operations were necessary to prevent increases in total dissolved solids, including the release of crude oil into the groundwater.

An article in the *Oil and Gas Journal* (Anonymous, 1957) dealt with brine disposal problems in the San Joaquin Valley in California. Contamination areas were under control through injection and percolation systems, but some oil operators were in the process of changing and correcting disposal techniques.

SIX

Groundwater Contamination Due to Agricultural Wastes

Stewart et al. (1967) analyzed core samples from fields and corrals in the South Platte Valley of Colorado to determine distribution of nitrates and water contaminants. High levels of nitrates were observed in cores sampled from irrigated fields. Low nitrate contents were determined in cores from irrigated alfalfa fields. The authors concluded that nitrates, under feedlot areas in this case, will not reach water table areas due to denitrification. However, nitrogen (e.g., as nitrate) detected under corrals ranged from 0 to over 5,000 pounds/acre. Large amounts of organic carbon and ammonia were concentrated in water samples beneath several corrals along with high bacterial counts. Groundwater contamination by deep percolation of nitrates occurred from corrals, but additional investigations were recommended to determine its full magnitude and complexity of movement.

Stewart et al. (1968) in another study, analyzed soil profiles and detailed the contributions of fertilizers and livestock feeding wastes to groundwater contamination in the same South Platte River Valley of Colorado. Nitrate levels in soil profiles varied widely with land use, and the data also showed that feedlots located near homesteads had a greater effect on nitrate content of domestic well water than a high concentration of cropland.

Robbins and Kriz (1969) presented a survey concerned with the relation of agriculture to groundwater contamination. In their study many agricultural sources of contamination were reviewed, including animal wastes, fertilizers, pesticides, plant residues, and saline wastewaters. Solutions to contamination and control problems were also included.

Biggar and Corey (1969) presented a comprehensive review of the relationship of agricultural drainage to water eutrophication. Attention was given to chemical reactions of nitrogen and phosphorus and their relationship to soil water systems. The fate of these nutrients transported by deep percolating water was analyzed. Illustrations showing plant nutrient loss from harvested areas and contributions of fertilizing elements from agricultural lands were also included.

Biggar and Corey (1969) presented a comprehensive review of the relationship of agricultural drainage to water eutrophication. Attention was given to chemical reactions of nitrogen and phosphorus and their relationship to soil water systems. The fate of these nutrients transported by deep percolating water was analyzed. Illustrations showing plant nutrient loss from harvested areas and contributions of fertilizing elements from agricultural lands were also included.

Viets (1971) focused on restricting fertilizer use because of leaching of nitrogen and phosphorus into groundwater. Results showed that groundwater contamination in agricultural areas also occurs from sewage, animal wastes, and various irrigation practices. Viets recommended sampling cores from surface areas on land down to water table depths and analyzing them for nitrate. This, he suggested, should be done before a specific fertilizer restriction is justified. He also maintained that restrictions on fertilizer use would not improve groundwater quality enough to compensate for the risk of a less abundant and more costly food supply.

Pesticides, Herbicides, and Groundwater Contamination

Bonde and Urone (1962) summarized data from a study of 225 wells in Adams County, Colorado. A plant toxicant (e.g., chlorate) was detected in the samples of one-quarter of the wells. These contaminated wells were located northwest of the Rocky Mountain Arsenal waste disposal basins and in the direction of groundwater flow. Chemical, X-ray, and bioassay techniques were used to identify the chlorate. High concentrations of sodium chloride were observed to coincide with the presence of other toxicants.

Studies have also been conducted on general adsorption and mobility characteristics of pesticides in soils. McCarty and King (1966; from U.S. EPA, 1974) reported that a significant correlation existed between the extent of adsorption and clay content of soils, and a correlation was apparent between adsorption and rate of pesticide movement. They emphasized that adsorption and degradation effects must be considered when predicting leachability of pesticides in soils.

Eye (1968) conducted research involving aqueous transport of dieldrin residues in soil. He concluded that the adsorptive capacity of the soil to dieldrin was so great that penetration through the soil was negligible. Thusly, no threat of groundwater contamination existed. A laboratory study on lindane and dieldrin on natural aquifer sands from Portage County, Wisconsin (Boucher and Lee, 1972) corroborated Eye's hypothesis (Eye, 1968). After three successive washes of distilled water, less than 20 percent of dieldrin adsorbed by aquifer sands was removed. However, 70 percent of adsorbed lindane had been leached.

Robertson and Kahn (1969) reported aldrin infiltrated through columns of Ottawa sand. Apparently, the penetrability of chlorinated hydrocarbon insecticides through soils depends upon the formulation applied, frequency of its application, soil conditions, and rate of rainfall or irrigation.

Dregne et al. (1969) studied the movement of 2,4-dichlorophenoxyacetic acid (2,4-D) in soils to determine the extent to which herbicides applied in the field enter groundwater systems. Primary emphasis was placed on the effect of cations on 2,4-D movement. A variety of analytical techniques indicated that 2,4-D in the salt or acidic forms was slightly adsorbed by soil particles. The ease of 2,4-D leaching was observed to depend on the permeability of the soils.

Johnson et al. (1967) reported on type and quality of insecticide materials observed in irrigated agricultural soils in the San Joaquin Valley of California. Small quantities of chlorinated hydrocarbon residues were detected in tile drainage effluent, but higher concentrations existed in open drains where subsurface drainage waters collected. Effluent samples from tile drains and tailwaters contained ten times the amount of insecticide residue. Large concentrations of residue were also observed in surface soils.

Schneider et al. (1970) described an experimental design concerned with movement and recovery of herbicides in the Ogallala aquifer at Bushland, Texas. Water from an irrigation well was used to inject common herbicides (e.g., picloram, atrazine, and trifluralin) into a well. The well was then pumped to recover recharged water. Water samples pumped from observation wells at radial distances of thirty and sixty-six feet from the previously sampled well showed that herbicides had moved through the aquifer with recharged water. Bacteria and some insecticides in this case were effectively adsorbed by fine Ogallala sand.

Swoboda et al. (1971) summarized research on toxaphene in Houston black clay in watersheds at Waco, Texas. Lewallen (1971) reported on pesticide contamination of a shallow well. Pesticide contaminated soil had been used as backfill around well casings. The contamination of the well water remained relatively low because of low solubilities of the pesticides DDT, DDE, and toxaphene in water. The contamination of the well seemed to have occurred through movement of surface soil containing adsorbed pesticides.

By the early 1980s, several incidents of groundwater contamination resulting from the field application of pesticides had been confirmed (Holder, 1986). The most widespread problems involved the insecticides/nematocides aldicarb (Temik) and DBCP (dibromochloropropane) (Rothschild et al., 1982). Early findings led to monitoring for other pesticides and several additional active ingredients have now been detected in groundwater in other states.

Groundwater contamination from field-applied pesticides had been unexpected. When the first incidents were documented, little pertinent baseline data on groundwater quality to assess the scope of emerging problems were available. When pesticides, therefore, were first detected in groundwater, no formal or informal regulatory mechanism was in place to respond. As a result, programs to detect pesticides in groundwater and to remedy problems of contamination were largely ad hoc. However, several patterns of response across states emerged. Opportunities to share knowledge and possibly coordinate activities occurred. In other cases, inconsistent decisions between states also indicated areas that needed research and reassessment.

Over two hundred pesticides are in common use, and a wide range of hydrogeologic conditions can affect the susceptibility of groundwater to contamination (NRC, 1986a). It is impracticable to sample every aquifer and test for every pesticide. Efficient monitoring, sampling, and testing and the collection of data that can be used to identify critical site/pesticide combinations are necessary. Residues in groundwater from field application of pesticides tend to be restricted to perched groundwater areas and the upper levels of unconfined aquifers. However, contamination may be more extensive, migrate more deeply, or reach confined aquifers, particularly when the source of the residues is more concentrated or the integrity of an aquifer-confining layer is pierced. Concentrated point sources of pesticides include manufacturing and formulating sites, applicator loading and mixing sites, and chemigation systems.

The emergence of the problem of pesticide residues in groundwater adds a new dimension to the whole array of public health, environmental protection, pesticide innovation and monitoring, and agricultural management (NRC, 1986a). Determining whether groundwater contamination could occur from the use of pesticides requires information on (U.S. EPA, 1986):

- What pesticides have the potential to leach to groundwater?
- What conditions are necessary or enhance the potential for a pesticide to leach to groundwater?
- What conditions other than "normal leaching" can result in groundwater contamination?

- Where has pesticide contamination occurred or is contamination a potential threat?

Concern for the exposures resulting from groundwater contamination by pesticides will require the previous information plus an understanding of (U.S. EPA, 1986; NRC, 1986):

- pesticide movement and persistence once they reach an aquifer
- who uses, and how frequently, groundwater supplies that are identified or predicted to be contaminated at levels of concern
- federal determination of health-based standards for pesticides in water
- the development of improved analytical tools for screening groundwater for pesticides
- development of a comprehensive systematic sampling program
- the development of improved models of pesticide behavior, fate in soils and aquifers
- the development of data on environmental fate and pesticide use patterns in conjunction with local hydrogeologic conditions, at a level to establish site-specific restrictions on their use
- the integration of groundwater resource conditions into an array of agricultural management practices and choices, including cropping, tillage, irrigation, and pest management.

Animal Wastes

Resnik and Rademacher (1970) reviewed the literature on causes and effects of animal waste runoff. Feedlots, in many instances, had been sited without regard to soil inventory and topographic characteristics. High BOD waste runoff and the infiltration of nitrates into wells from manures was documented. The extent of the problem and status of regulatory legislation to 1970 were also discussed.

Since the extent of groundwater contamination problems caused by animal wastes has been a strong concern, substantial work has also been conducted on methods of control. For example, Loehr (1970) analyzed liquid and solid effluents from anaerobic lagoons that treated feedlot wastes. Under ideal equilibrium conditions, the liquid effluent from such lagoons constituted a groundwater contamination threat. However, when used as a treatment unit, anaerobic lagoons were effective for treating livestock and feedlot wastes that showed high solids content. Data on quality and quantity of runoff from cattle feedlots were also presented by Loehr (1970). Minimum drainage control in this instance was possible using retention ponds because of the intermittent nature of this runoff. Resulting groundwater contamination problems via use of these methods were also discussed.

Other studies have also been conducted on specific animal

waste contamination problems. Stewart et al. (1967), reported on nitrate contamination of groundwater in the South Platte Valley of Colorado, an area intensively farmed and one with many livestock feeding operations concentrated together. The average total nitrate–nitrogen content in soil profiles for various kinds of land use for this area was reported by Stewart et al. (1967). Groundwater samples also often contained high concentrations of nitrate and those obtained beneath feedlots contained ammonium nitrogen and organic carbon. In addition, the data summaries showed that nitrates were moving towards groundwater supply areas from feedlots and most irrigated fields, with the exception of those planted with alfalfa crops.

Gilbertson et al. (1971) described runoff, solid wastes, and nitrate movement characteristics on beef feedlots in Nebraska. Runoff quality and quantity were observed to depend on rainfall more than on slope or cattle density. High density lots yielded 150 percent more winter runoff than low density lots and after one year, nitrate movement in soil was minimal.

Lorimer et al. (1972) reported nitrate concentration in groundwater areas beneath cattle feedlots in Central City, Nebraska. Sampling of wells near feedlots revealed that the beginning of irrigation pumping resulted in no significant increase in nitrate levels. The levels were observed to be below the U.S. PHS limit.

Gillham and Webber (1968, 1969) examined nitrogen contamination of groundwater by barnyard leachates. Quantitative flow nets were drawn for the area and this permitted discharge calculations. The concentration of nitrogen was related to direction of groundwater flow, the presence of conditions suitable for leaching, and the actual dilution rates of the local groundwater system.

Miller (1971) reported field and laboratory studies on infiltration rates, nitrate distribution, and groundwater quality parameters beneath cattle feedlot areas in the Texas High Plains. Feedlot wastes entering the water table below these areas were not detectable in most localities. Infiltration of feedlot runoff and concentration of dissolved ions in groundwater, however, were dependent on geology, depth to the water table, and differences in lateral and vertical permeabilities of the underground formation.

Crosby et al. (1971) analyzed drilling at a dairy in the Spokane Valley of Washington to study the effects of feedlot operations on groundwater quality. Fecal coliforms were isolated from samples taken within a few feet of the ground surface. Chlorides and nitrates were observed at depths that could reach groundwater. Low moisture content of the soil indicated that it had stabilized.

Irrigation Return Flows and Groundwater Contamination

Several studies on the effects of irrigation on soils and groundwater have been conducted in Oahu, Hawaii (U.S.

EPA, 1974), although determining such impact is not an easy task (Khan, 1980). Mink (1962) reported an increase of silica and nitrate in the groundwater beneath heavily irrigated sugar cane. Contamination of the groundwater was due to nitrate fertilizer and silica leaching through water-logged soil.

Leonard (1970; from U.S. EPA, 1974) discussed the effects of irrigation on the chemical quality of ground and surface water in the Cedar Bluff Irrigation District of west-central Kansas. One hundred observation wells were established as monitoring stations. The chemical quality of the ground-water from these areas was varied between wells. Calcium, sulfate, and bicarbonate ions were abundant and the chlo-ride content increased as the irrigation season continued. It appeared that the groundwater in the district was being diluted and displaced by irrigation water.

In arid areas, more irrigation water than is needed for evapotranspiration must be applied to the soil to avoid accu-mulation of salts in the root zone (Bouwer, 1987). This creates a downward flow of water from the root zone to the underlying groundwater. The resulting deep percolation water not only contains the salts that were in the irrigation water, but also fertilizer and pesticide residues that are a potential source of contamination for the groundwater. The potential for groundwater contamination under such condi-tions is extensive and serious. Soluble salts, nitrates, and pesticides, are the chemicals in deep percolated water that are of great concern (Schmidt and Sherman, 1987). The results of studies indicate that irrigation return flow usually exerts a substantial impact on groundwater quality. In addi-tion, extensive pollution of shallow groundwater in parts of the San Joaquin Valley have been caused by the use of the pesticide DBCP (Schmidt and Sherman, 1987). Other studies (Sabol et al., 1987) indicate that the groundwater quality in Arizona and New Mexico have been deleteriously affected in deep aquifers and in shallow aquifers that are hydraulically connected to surface water supplies. The mag-nitude and time rate of groundwater quality changing as a function of the following: irrigation management practice; fertilizer and pesticide applications; quality of irrigation water; rate of groundwater level decline; presence of perched zones that intercept percolating water; proximity to surface water supplies; leakage through and along well casings; and, finally, because of soil salinity. Irrigation effects in Oklahoma and Texas have also been reviewed (Law, 1987). More efficient irrigation and fertilizer manage-ment practices are required to minimize adverse impacts on groundwater.

SEVEN

Contamination of Groundwater from Disposal and Injection Wells

Groundwater contamination problems related to subsurface disposal of liquid wastes by deep-well injection have been reviewed in the literature since 1950 (U.S. EPA, 1974).

Reck and Simmons (1952) reported on groundwater in the Buffalo-Niagra Falls region of New York. They described that the quality was good. However, large sections of the Onondaga Limestone aquifer had been contaminated by individual and industrial wells that had been drilled specifically for injection of waste materials. The wells had become clogged and lost their capacity to absorb waste.

For example, Warner (1965) in his studies, concluded that deep-well injection was technically feasible and if properly implemented was considered a safe method for disposal of liquid wastes. Walker and Stewart (1968) and Talbot (1968) reviewed deep-well disposal practices and regulations for various states. Their findings mention that a suitable disposal stratum and a waste physically and chemically compatible with the resident material in the disposal formation is necessary. Caswell (1970), however, when reviewing the technology, hydrology, and legal status of deep disposal wells, warned that in some cases injection was not feasible if wastes were long-lived or the hydrogeology of the disposal horizon was not satisfactory. Rinne (1970) stressed the need for detailed geologic–hydrologic investigations and comprehensive studies on the suitability of various sites. Proper system design was also of great importance.

Articles in the *Water Well Journal* (Anonymous, 1968a) and *Environmental Science and Technology* (Anonymous, 1969) also summarized data on 110 injection wells in use throughout north-central and Gulf Coast areas to 1969. The characteristics of industrial waste disposal wells reviewed in these articles included: operations, locations, well depths, depth of injection horizons, geologic horizons, formations, chemical and physical character of wastes, and injection pressure and rates. Problems of design, control, and monitoring of deep-well injection systems were also examined during this same time period.

In a 1966 laboratory study, Warner (1966) compared reaction rates between injected and interstitial solutions to the dispersive character of porous medium. The concept that a buffer zone of nonreactive water should be established between injected waste and aquifer water had been proposed.

Basic design principles for disposal well systems were presented by Marsh (1968), Slagle and Stogner (1969), and McLean (1969). Design requirements that allowed aquifer protection strategies included: selection of a zone bounded by aquicludes, strict drilling, casing, and sealing procedures, waste quality and application rate controls, proper surface injection and treatment equipment, use of standby wells, knowledge of hydraulic gradients and hydrodynamic dispersion factors, and a comprehensive system of monitoring wells. McLean (1969) also emphasized the need for pre-construction testing of the sites and testing of the formation's hydrologic properties were urged. A method for calculating the radius of injection capacity of formations was also described.

Effects of deep injection are complex and geologists feel that in some instances the use of injection wells is misunderstood or misapplied. Sheldrick (1969; from U.S. EPA, 1974) summarized pertinent criticisms of geologic criteria on which to evaluate the feasibility and safety of injection wells. Knowledge of the hydrodynamics of underground formations and use of monitoring techniques were thought to be inadequate for waste injection at that time.

Tofflemire and Brezner (1971) summarized existing deep-well injection practices in the United States through 1971, with particular reference to the New York area. Selection of sites, well construction, waste quality, and well-monitoring criteria were reported. Salt water, industrial, and radioactive wastes were the major types of liquid categories discussed in reference to deep-well disposal.

Clearly and Warner (1970) presented an extensive monograph on underground wastewater disposal for the Ohio

River Basin in 1969. Insight into public policy issues was provided in this report. Administrative and regulatory guidelines were also described. This was done in an effort to aid in the evaluating, the location, design, construction, operation, and abandonment of injection wells. At this time the Ohio Valley was considered amenable to waste injection. It was also, however, recommended that limited quantities of wastes be regarded as eligible for subsurface disposal and that monitoring needs were important.

Bergstrom (1968) studied the feasibility of industrial waste disposal by injection wells in Illinois and reviewed the suitability of various geologic formations to accommodate such use. Geohydrologic conditions made disposal by injection wells feasible in much of the southern two-thirds of Illinois. However, testing, evaluating and accepting site conditions were still considered necessary before any well installations could be approved. The basic design policies incorporated into this type of industrial waste disposal permit system were described by Smith (1971).

A study by Jones (1947; from U.S. EPA, 1974) of injection wells for the subsurface disposal of oil field brines in Kansas emphasized protection afforded groundwater supplies and the need to repressure "played-out" oil fields. The study also warned of the battle that occurs with corrosion of brine handling equipment in the use of injection wells.

Grubbs et al. (1970; from U.S. EPA, 1974) evaluated geologic and engineering parameters governing disposal of liquid wastes into deep-well injection areas in Alabama. Their study included an evaluation of geological and hydrological factors to identify favorable subsurface reservoirs for waste confinement. A design and cost procedure was also provided concerning rapid feasibility studies of proposed sites. An extensive bibliography on liquid wastes was also included in their report. Tucker (1971) also reviewed operations of injection wells over a six-year period in Alabama. He described well design criteria and well monitoring techniques through that time period.

Dean (1965), Barraclough (1966), and Goolsby (1971) reported on the use of a deep injection well system operated by Chemstrand Company at Pensacola, Florida, for the disposal of aqueous process wastes from the manufacture of nylon. The wastes were injected at low pressures and high rates into the lower limestone of the Floridan aquifer between two thick beds of clay. Dean (1965) described design criteria (i.e., the casing program), construction, and operation of this system. Goolsby (1971) used monitoring wells to show that the waste extended outward about one mile from the injection wells and pressure effects from the system extended outward over twenty-five miles.

Garcia-Bengochea and Vernon (1970) described deep-well disposal of industrial wastes in high saline boulder zones of the Floridan aquifer in Florida. The potential of the saline boulder zone for similar uses, if hydrogeological conditions were typical of the region, was also discussed. Kaufman (1973) in another study analyzed data on deep-well injection of industrial and municipal effluents and concluded that injected wastes remained confined in receiving strata, at least in the northwestern Florida area. The crucial need for monitoring of hydraulic and geochemical effects was emphasized.

Henkel (1953) reported the development of a waste disposal system at the DuPont Company's adiponitrile plant near Victoria, Texas. Concentrated liquid chemical wastes were being injected into deep wells. Construction procedures (i.e., cementing and anti-corrosive casing) for well development and brine treatment operations (e.g., aeration, induced precipitation, filtration, and chlorination) were described. Veir (1969) reported on deep-well disposal systems in Bay City and Odessa, Texas. Geologic and hydrologic sampling programs and techniques were discussed. The details of well construction, the equipment used, and waste pretreatment processes were also summarized.

Evans (1966) and Evans and Bradford (1969) examined the possible relationship between area earthquakes in Denver, Colorado and deep injection wells at the Rocky Mountain Arsenal which was disposing of wastes from the manufacture of toxic gases. The reports mentioned the unknown dangers that could occur. Evans and Bradford (1969) also warned that deep injection well techniques offered temporary and not long-term safety from the permanent toxic wastes injected.

Recharge Wells

The University of California Sanitary Engineering Research Laboratory (Krone et al., 1957) conducted a study of contamination travel from direct well recharge. The study area included a well which consisted of a recharge well plus twenty-three observation wells penetrating a confined aquifer 100 feet underground. Freshwater and settled sewage were injected at various rates into the well. Chemical, bacteriological, and radiological tracking techniques were used to determine rates of travel of recharge water. Bacterial contaminants travelled 100 feet in the direction of groundwater movement, although steep gradients occurred. Well clogging was also examined in this study.

Wesner and Baier (1970) reported on their research conducted within the Orange County Water District in California. Their investigation involved wastewater reclamation and subsurface injection. Their objectives included:

- the study of hydraulic characteristics of the injection barrier system of multi-point injection wells
- the study of long-term fates of reclaimed wastewater in injection systems
- the study of the feasibility of utilizing wastewater as a barrier
- the study of the chemical composition of blended reclaimed water derived from deep groundwater sources

The performance of the subsurface injection system was satisfactory, but persistent odors and tastes in injected reclaimed water was a serious deterrent to utilizing this type of method to dispose of wastes.

Studies have also been conducted to reclaim water from effluent of Nassau County, Long Island, New York (Cohen and Durfor, 1966; Baffa and Bartilucci, 1967; Anonymous, 1968b; Peters and Rose, 1968; Rose, 1968, 1970). These studies also dealt with alternative methods of waste renovation, comparisons of recharge wells with recharge basins and area withdrawal rates.

A field study of water quality in the municipal well of Aron, South Dakota, was reported by Jorgensen (1968). The water quality from the municipal well and from nearby wells tapping the same aquifer were different. Additional analyses of water samples from these different wells showed the anomaly to be caused by leakage from a nearby abandoned well which tapped an adjacent aquifer.

Ham (1971; from U.S. EPA, 1974) surveyed the problems concerned with contamination and well pumping. He presented information on the various sources of contamination that could affect wells. Eleven items dealing with statutory and administrative control measures were also included. Jones (1971), in a similar review concluded that improperly built and abandoned wells serve as unauthorized and uncontrolled recharge points and degrade groundwater quality.

Industrial Injection Wells

Wells for subsurface disposal of industrial wastes were implemented in the 1930s (U.S. EPA, 1973). However, the extent of their use was limited until the 1960s when increasing emphasis on water contamination control caused industrial concerns to seek other alternatives for wastewater disposal. Subsurface injection was one of the alternatives chosen and as of mid-1972, 246 wells had been constructed in the United States (Warner, 1972; from U.S. EPA, 1973).

At the same time, considerable concern was being expressed about the use of injection wells, primarily because of the following:

- Some of the wastewaters that were being injected contained toxic chemicals that have an indefinite persistence rate in the subsurface environment.
- Monitoring groundwater environments was difficult, especially when compared to surveillance of surface areas.
- If contaminated groundwater was used for water supply or other essential needs, decontamination may be difficult or impossible to achieve.

Inventory of Wastewater Injection Wells

An industrial injection well inventory conducted in the United States in the early 1960s listed thirty wells (Donaldson, 1964). Subsequent inventories by Warner (1967, 1972) listed 110 and 246 wells, respectively. Only twenty-two wells had been constructed before 1960, although twice that number existed in 1964. Significant information related to wells constructed during this time period included that (U.S. EPA, 1973):

- A high percentage of wells were constructed by chemical, petrochemical or pharmaceutical companies, and a smaller percent by refineries and natural gas plants.
- Many of the wells constructed were operated for some time.
- Existing wells at depths between 2000 and 6000 feet tended to provide additional assurance of protection to usable groundwater resources.
- Many of the wells constructed were injected with less than 200 gallons per minute.
- A low percentage of the existing wells at that time were injecting fluids at well-head pressures exceeding 1,500 psi.

Environmental Consequences of Wastewater Injection

Wastewater injection can modify groundwater systems and introduce into subsurface areas liquids with a different chemical composition than the present natural fluids (U.S. EPA, 1973). Injection degrades high quality groundwater, contaminates other resource areas, such as petroleum or coal reserves, stimulates earthquakes, and causes chemical reactions to occur between wastewater/natural water. Rock material can also react with the wastewater during injection intervals. These impacts are predictable and can be quantified in each individual situation. In the case of existing wells where significant adverse environmental effects are not expected to occur, regulatory authorities can permit the sites. Conversely, where untenable impacts are anticipated, permits can also be denied.

Escape of wastewater through a well bore can occur into a freshwater aquifer because of insufficient casing, corrosion of the casing or by failure of it. Vertical escape of injected wastewater can also occur outside the well casing area from an initial injection zone into a freshwater aquifer and through confining beds that are inadequate because of high primary permeability, solution channels, joints, faults or induced fractures. Injected wastewater can then escape from the injection zone through nearby deep wells that have been improperly cemented or plugged or that have been completed with insufficient or corroded casing. Direct contamination of fresh groundwater occurs then by lateral movement of injected wastewater (i.e., from a region of saline water) in the same aquifer. The distances involved and low rates of travel of wastewater make the probability of direct contamination small, but it can be significant on the long-term basis.

Indirect contamination of fresh groundwater can also occur when injected wastewater displaces saline water, causing it to flow into a freshwater aquifer zone. Vertical flow of saline water in this instance occurs through paths of natural, induced permeability areas in confining beds or through inadequately cased or plugged deep wells occupying the same area. If large volumes of wastewater are injected near freshwater/saline water interfaces, a common occurrence in many coastal aquifers, and inland locations, fresh groundwater interfaces can be displaced with saline water.

Geologic and hydrologic circumstances in which earthquakes are stimulated by wastewater injection are not fully understood. However, a fault along which movement is subject to fluid injection will release strained energy and cause earthquakes (U.S. EPA, 1973). Among industrial injection wells surveyed by Warner (1972), few were present in such locations, and other than the Rocky Mountain Arsenal well near Denver, none of them had been related to earthquakes.

CONTROL OF INDUSTRIAL WASTEWATER INJECTION INTO SALINE AQUIFERS

Processes, procedures, and methods for controlling industrial wastewater injection into saline aquifers described by U.S. EPA (1973) included:

- an extensive evaluation of hydrogeological frameworks and restrictions on regionally unsuitable locations and aquifers for wastewater injection
- specific investigations of the local hydrogeological environment and restricting locally unsuitable locations and aquifers from receiving wastewater injection
- evaluating specific fluids injected into aquifers as wastewater
- the need to estimate the extent of chemical reactions between injected fluids and aquifer fluids, minerals, and any heat generated
- evaluation of suitable construction features for properly locating wells
- continuing of hydrogeologic evaluations during construction and testing of wells
- determining basic aquifer characteristics and measurements of actual aquifer response to injected wastes, including rate of movement of fluids
- detailed operating programs for injection well usage
- establishment of essential surface equipment and programs for emergency procedures, including rapid shut-off systems, standby emergency facilities (i.e., including long-term decontamination areas)
- the need to establish proper abandonment procedures and follow-up for all wells

- implementation of monitoring programs for injection wells after closure
- systematic long-term monitoring programs for aquifers located in the area of injection wells

Regional Geological Considerations of Injection Sites

Specific injection well sites must include a detailed analysis of local geology. Just as some basins are geologically favorable sites for deep-well injection, other areas are unfavorable because sedimentary rock cover is thin or absent. Extensive areas where impermeable igneous-intrusive and metamorphic rocks are exposed at the surface can be eliminated from consideration for waste injection (U.S. EPA, 1973). The exposure of igneous and metamorphic rocks are significant because the sedimentary sequence extends toward them and the salinity of the formation water decreases toward the outcrops around the exposures. Where thick volcanic sequences lie at the surface, they also are not suitable for waste-injection wells. Volcanic rocks have fissures, fractures, and interbedded gravel that accept injected fluids. However, they may contain reserves of freshwater that can also be contaminated.

Geological areas not underlain by major basins or prominent geologic features may be satisfactory for waste injection if they are underlain by a substantial amount of sedimentary rocks that contain saline water. This is also the case if the potential injection zones are sealed from freshwater-bearing strata by experimentally verified impermeable confining beds (U.S. EPA, 1973).

Local Site Evaluation

The factors that must be considered when evaluating waste injection well sites according to U.S. EPA (1973) also include:

- conducting regional geologic and hydrologic framework evaluations, including structural geology, stratigraphic geology, groundwater geology, mineral resources, seismicity, hydrodynamics
- considering local geology and geohydrology, such as structural geology, geologic description of sedimentary rock units, lithology, description of potential injection horizons and confining beds, thickness and vertical and lateral distribution characteristics, porosity, permeability, chemical characteristics of reservoir fluids, groundwater aquifers at the site and in the vicinity, thickness, general character, amount of use and potential for use, mineral resources and their occurrence at well site and in immediate area, oil and gas, coal, and brines

Vertical confinement of injected wastes is important for the protection of water resources and for the development of deposits of oil, gas, other minerals and resources. Waste injection into limestone and dolomite has been successful in some areas because the permeability of these rocks improves with acid treatment (U.S. EPA, 1973). Unfractured shale, clay, slate, anhydrite, gypsum, marl, and bentonite have also been observed to provide good seals against the upward flow of fluids. Limestone and dolomite may be classified as satisfactory confining strata. However, these rocks also contain fractures and solution channels. Their adequacy as injection sites must be determined for each individual case.

Minimum depths of burial, the necessary thickness of confining strata, salinity concentrations in the injection zone, have not always been quantitatively described, and full constraints therefore are not always predictable for a site. The depth of the injection zone extends to where a confined saline-water-bearing zone is present and it may range from a hundred to many thousand feet. Minimum salinity concentrations of water should be at least 1,000 mg/l of dissolved solids (U.S. EPA, 1973).

The minimum thickness and permeability necessary to allow fluid injection at desired rates can be estimated from equations developed by petroleum engineers and groundwater hydrologists (U.S. EPA, 1973). These aquifer hydrodynamic equations are significant when evaluating waste-injection well sites. Natural hydrodynamic gradients in the injection zones cause injected wastes to be distributed asymmetrically about the well bore and transported throughout the aquifer even after injection has ceased. Hydrodynamic dispersion (i.e., the mixing of displacing and displaced fluids during movement through porous media) also causes a wider distribution of waste material within the injection zone. Dispersion in this instance could lead to rapid lateral distribution of waste in heterogeneous sandstone and fractured strata, although sorption of waste constituents by aquifer minerals can retard the spread of waste from initial injection sites.

Mathematical models are now also available that satisfactorily predict movement of waste in aquifers under restrictive, simplified physical circumstances (U.S. EPA, 1973). These models, however, do not always present quantitative estimates of the rate and direction of movement of injected wastes.

Maximum pressure at which injection occurs without causing fracturing may also limit intake rates and the actual operating life of an injection well system (Hubbert and Willis, 1957). Injection pressures at which hydraulic fracturing occur relate directly to rock stresses and strength of the injected zone. Pressures at which fracturing occurs should be estimated before drilling, which can be done from experience gained in drilling nearby oil fields.

Other considerations for determination of sites suitable for injection of waste into subsurface areas include (U.S. EPA, 1973):

- an evaluation of high fluid pressures and temperatures in the anticipated injection zone area that could make the process difficult or uneconomical
- an evaluation of the incidence of earthquakes in the area that caused movement along faultlines and also damaged subsurface well facilities
- the presence of abandoned, improperly plugged wells that also penetrate the same injection zone to be used (These abandoned wells provide a means for escape of injected waste to groundwater aquifers or to the surface.)
- an extensive geological evaluation of the injection zone, the chemistry of interstitial waters
- an evaluation of the possibility that fluids injected into an unstable area may contribute to the occurrence of earthquakes

Wastewater Injection and Feasibility

When evaluating the feasibility of injecting wastewater, the character of the untreated material must be determined (U.S. EPA, 1973). Volume, physical characteristics (e.g., specific gravity, temperature, suspended solids content, gas content, etc.), chemical characteristics (e.g., various inorganics, organics, pH, chemical stability, reactivity, toxicity, etc.), and biological considerations must be taken into account. The suitability of any type of waste for subsurface injection then basically depends upon its volume and physical–chemical properties.

Disposal of wastes into subsurface aquifers also constitutes use of limited storage space. Fluids injected into deep aquifers displace the same amount of liquid that is injected and eventually saturate storage zones. Since intake rates of injection wells are also limited and their operating life depends on the total quantity of fluid injected, injection pressures can be limiting because excessive amounts will cause fracturing and damage the confining strata (U.S. EPA, 1973).

Maximum injection pressures in many states are specified by regulatory agencies, which seldom allow pressures over 0.8 psi/ft. (U.S. EPA, 1973). However, initial pressures required to inject waste at specified rates can be computed, if physical properties of the aquifer and the injected waste are known.

Since the injectability of specific wastes depends on physical and chemical characteristics of the waste, physical and chemical characteristics of the aquifer, and native aquifer fluids, physical and chemical interactions between the injected waste and the aquifer fluids and characteristics affect plugging of the aquifer pores, and consequently, loss of intake capacity occurs (U.S. EPA, 1973). Plugging can also be caused by the presence of suspended solids or gas in the injected waste. Reactions between injected fluids and aquifer minerals and between injected and interstitial fluids also cause plugging. Bacteria and mold also plug areas near the well bore. Knowledge of the mineralogy of the aquifer and the chemistry of interstitial fluids and wastes should indicate what type of reactions can be anticipated during the injection process. Laboratory tests can also be conducted with rock cores and wastewater injection samples if confirmation of anticipated reactions are necessary.

Selm and Hulse (1959; from U.S. EPA, 1973) described the reactions occurring between injected and interstitial fluids that form plugging precipitates. These reactions include:

(1) Precipitation of alkaline metals such as calcium, barium, strontium, and magnesium

(2) Precipitation of insoluble carbonates, sulfates, ortho-phosphates, fluorides, hydroxides, ortho-phosphates, and sulfides

(3) Precipitation of oxidation–reduction reaction products

If plugging is considered to be a possibility, the waste can be treated to make it non-reactive. Non-reactive water can then also be applied before waste is injected. This is done to form a buffer in the aquifer (Warner, 1966).

Other minerals that react with injected wastes include acid-soluble carbonate and clay minerals (U.S. EPA, 1973), although acidizing of reservoirs containing carbonate minerals is expected to be beneficial. An undesirable effect of the reaction, however, involves evolution of carbon dioxide, which in turn increases pressure and causes plugging, if present in excess of its solubility products.

Ostroff (1965) and Warner (1965, 1966) have also presented written summaries concerning injection of wastes. Factors that bear on injection, such as aquifer mineralogy, temperature–pressure, and chemical quality of aquifer fluids, should be a part of any feasibility report. The treatment necessary to make a specific waste injectable and compatible with the environment should be part of any waste management program.

Evaluation of Injection Wells and Their Construction

Geologic situations vary and the different characteristics of waste preclude establishment of exact specifications for construction of injection wells. Each injection system needs to be considered on its waste volume and type. Geologic and hydrologic conditions that exist must also be considered. General specifications, however, can be summarized (U.S. EPA, 1973).

Construction of well facilities for establishment of waste injection systems include drilling, logging, testing, and completion activities. Holes must first be drilled, logged, and tested before they are completed as injection wells. Completion phases should include installation and cementing of the casing, installation of injection tubing, and perforating the casing that stimulates the injection horizon. It is also necessary to install and cement some of the casing during drilling.

Drilling programs should also be designed to permit installation of necessary casing strings with sufficient space around casing areas for cement. Samples of the rock formations penetrated should be obtained during drilling and water samples collected at different horizons to provide essential geologic and hydrologic data. Complete logging and testing of wells intended for injection of wastes should be conducted, and pertinent data filed with appropriate agencies. The information desired involves porosity, permeability, fluid pressure in formations, water samples, geologic formations intersected by bore holes, thickness and character of disposal horizons, mineral content of formations, temperature of formations, and amount of flow into various horizons.

Design of a casing program depends on well depth, character of rock sequence, fluid pressures, type of well completion, and the corrosiveness of fluids that contact the casing (U.S. EPA, 1973). A casing string is installed below the base of the deepest groundwater aquifer where fresh groundwater supplies are present. Small diameter casing strings are then set just above the injection horizon, depending on whether the hole is to be completed or to be cased and perforated. The annulus between rock strata and casing is then filled with cement grout. This grouting is added to protect it from corrosion, to increase strength, to prevent mixing of the waters contained in the aquifers behind the casing, and to forestall travel of injected waste into other aquifers. Wastes should be injected through separate interior tubes rather than being in contact with the well casing. This is especially important when corrosive wastes are being injected.

If it is desired to increase the acceptance rate of injection wells by chemical or mechanical treatment of the injected zone, attention should also be given to stimulation techniques (e.g., hydraulic fracturing, perforating, and acidizing) to insure desired intervals are treated and no damage to the casing, cement, or confining beds occurs (U.S. EPA, 1973).

Wastewater Movement in Receiving Aquifers and Aquifer Response

Estimates of pressure buildup in the receiving aquifer are important because the maximum at which liquids can be injected may limit the intake rates and operating life of actual injection wells (U.S. EPA, 1973).

Estimates can be made on increase of pressure in the receiving aquifer and projected wastewater injection rates.

Van Everdingen (1968; from U.S. EPA, 1973) outlined a methodology for estimating pressure buildups resulting from injection wells.

Estimating lateral movement of injected wastewater is needed so location of underground spaces occupied by the discharge can be used to regulate and manage the subsurface release. The extent and direction of wastewater movement should be made after geohydrologic characteristics of the receiving aquifer have been determined. These estimates are difficult and may be modified by natural flow systems in specific aquifers. Hydrodynamic dispersion, porosity and permeability variations, density and viscosity differences between injected and interstitial fluids, also add to the complexity of any evaluation.

Operating Program

The operating programs for waste injection systems should conform to the geological and engineering properties of the injected horizon (U.S. EPA, 1973). Complete volume and chemistry of the waste fluids must also be considered. Injection rates and pressures must be evaluated, since pressures depend on volume injected. Pressures are limited to values that prevent damage to the injection facilities or confining formations. Operation schedules involving rapid variations in injection rates, pressures, or waste can damage the facilities. Therefore provisions should be made for shut-off in the event hazardous flow rates, pressures or waste quality fluctuate beyond acceptable limits.

Emergency Procedures and Surface Equipment for Injection Systems

Surface equipment for wastewater injection includes holding tanks, flow lines, filters, pumps, monitoring devices, and standby facilities (U.S. EPA, 1973). Equipment associated with injection wells should be compatible with waste volumes, including physical and chemical properties of the waste. The equipment should insure that the system operates as efficiently as possible. Experience with injection systems in the past has revealed difficulties due to improperly selected equipment and corrosion of pumps. Equipment should also include adequate pressure and volume monitoring equipment. If injection tubing is used, pressures should be monitored. Fluid pressures in the tubing and in the annulus between it and the casing should be included. This pressure and volume monitoring is a means of detecting tubing or packer leaks. Automatic alarm systems can also signal the failure of any important component of the injection system. Filters should also be equipped to detect effluents with high suspended solids content. Chemicals that have previously been determined to be deleterious to the injection program should also be monitored.

Standby facilities are essential to cope with malfunction of a well and provisions should be made for alternative waste facilities in the event of injection system failures. Alternative facilities include use of standby wells, holding tanks, or an on-site treatment plant. Additional facilities and procedures should be available for use in engineering failures of the system or recognizing that a subsurface area is being contaminated. Handling of corrosive wastewater would be sufficient for advance planning to be used if tubings fail (U.S. EPA, 1973). Corrective procedures should begin immediately, and injection of non-corrosive liquids into the wells should commence until the well bore is completely cleared. The well should then be shut down until reservoir pressures have decreased to a level allowing removal of the damaged tubing.

Monitoring Procedures for Injection Wells

Monitoring of injection systems or within injection zones, should be a continuous function. Monitoring should include measurement of the physical, chemical, and biological characters of injected fluids. Periodic checking of casings and tubing for corrosion, scaling or other defects should also be conducted. Reasons for monitoring injection zones or adjacent aquifers include the necessity to determine specific fluid pressures, including rate and direction of movement of injected wastewater and the sum total of aquifer fluids. However, monitoring with wells to determine rate and extent of movement of wastewater within injection zones is limited because intercepting wastewater is difficult and information obtained is difficult to interpret (Warner, 1965).

A more feasible approach to monitoring wells is to evaluate the fluid pressure in the injection zone or adjacent aquifers (U.S. EPA, 1973). Monitoring wells can also be constructed for this purpose. Goolsby (1971) discusses examples of injection systems where these types of wells were used for detection of waste travel and measurement of reservoir fluid pressures.

Another type of monitoring well that can be used in conjunction with wastewater injection systems is one constructed within freshwater aquifers and is located near the injection well. These wells provide a means for detecting leakage from the specific injection well in question, i.e., contaminants entering the supply aquifer tend to move toward a discharging well (U.S. EPA, 1973). The quality of water from nearby springs, streams, and lakes should also be monitored to detect possible effects that the waste disposal injection well operation might be having.

Injection Well Programs at the State Level

Regulation of injection wells at the state level varies. States that produce oil regulate the disposal of oilfield water through an oil and gas agency, but other categories of disposal wells are frequently regulated through water pollution control activities, environmental protection legislation or health regulations. Some states have developed specific laws, regulations, and policies concerning industrial wastewater injection, with Texas being the first to establish laws, specifically, some concerning industrial wastewater injection wells. Since then, other states have passed legislation to include underground injection.

ORSANCO, in 1973 (U.S. EPA, 1973), incorporated eight steps for use by member states (Illinois, Indiana, Kentucky, New York, Ohio, Pennsylvania, Virginia, and West Virginia) for injection of wastes. The steps incorporated included:

- Applicant must assess the geology and geohydrology at the proposed well site and determine the suitability of wastewater for injection, with studies to be made in consultation with state agencies.
- There is an obligation to apply to the specific state agency that has legal jurisdiction for permits to drill and test wells for subsurface wastewater injection. The application must be supported by reports that document details of the proposed injection system, including monitoring and emergency standby facilities; and on issuance of permission, the applicant will be informed of geologic and geohydrologic parameters to be employed by the state in its deter-

mination on feasibility of injecting wastewater into the well.
- Drilling of the well starts and submitting logs, test information, including a well-completion report.
- The applicant requesting approval to inject wastewater into the well should indicate changes, if any, from the original plan.
- The state should conduct a prompt evaluation and the systems approved as is, with modification or disapproved based on geologic, hydrologic, or engineering data submitted for evaluation.
- If approved, operation of the injection system should transpire in accordance with state requirements. Appropriate regulatory agencies should be notified if problems arise, if remedial work is required, or if significant changes in the wastewater stream occur or are anticipated.
- Appropriate agencies in adjacent state(s) should be provided the opportunity to review and comment on applications where a proposed injection well is to be located within five miles of their state borders. These agencies should also be advised of any problems occurring during the operation of each well.
- The eventual well abandonment will be in accordance with state regulations or other technically acceptable procedures.

EIGHT

Other Wells as Sources of Contamination

In addition to the types of industrial wastewater injection wells previously discussed, other deep wells are also sources of groundwater contamination. Such wells include those completed in association with oil production and exploration activities, mining, geothermal energy production, sewage treatment, desalination, radioactive waste disposal, and underground gas storage.

Wells in the Petroleum Industry

Deep wells are used by the petroleum industry for exploration, production of oil and gas, and for injection of brines back into subsurface areas. Brine injection may also in some instances, be used to maintain reservoir pressure, provide a displacing agent in secondary recovery of oil, or be used to dispose of brine wastes.

The number of petroleum exploration and production wells drilled in the United States by various sources is in the millions. Iglehart (1972) reported that 27,300 wells were drilled in 1971, a year in which activity was low. The exact number of brine injection wells drilled is not known, but oil producing states have reported that in 1965, 20,000 wells were recorded in Texas alone (Warner, 1965). Probably an equal number were also drilled throughout the other states. Information gathered by the Interstate Oil Compact Commission (1964; from U.S. EPA, 1973) indicated that 360 billion gallons of water were pumped in 1963 in conjunction with petroleum development or production, although 72 percent of the produced water was reinjected.

Recognizable hazards to groundwater may result from deep wells, including oil and gas petroleum production wells if they are inadequately cased, cemented, plugged, or abandoned. These wells can also provide inter-aquifer movement of saline groundwater. The danger to usable groundwater therefore exists because of the hundreds of thousands of oil and gas wells drilled during the last 100 years that were abandoned, with many of them not being properly plugged. Groundwater contamination caused by improperly abandoned oil and gas wells has been described for some petroleum producing states (Fryberger, 1972; Wilmoth, 1971; Thompson, 1972).

Groundwater contamination from oilfield brine injection wells is essentially the same as industrial wastewater injection wells. However, since oilfield brine is considered natural water, it normally does not contain chemicals that are toxic. However, the high levels of dissolved solids that occur in most brines and the substantial volumes involved, present a situation that can degrade large amounts of usable groundwater (Ostroff, 1965).

A detailed investigation of an incident of contamination of a freshwater aquifer by oilfield brines was conducted by Fryberger (1972). He described the extent of a brine contamination incident as including one square mile which in time spread and affected four-and-one-half square miles. This contamination incident is predicted to remain for over 250 years in the aquifer before being flushed out naturally.

Wells and Solution Mining

Wells have been used to extract sulfur, salt, and other minerals from subsurface areas by injection of water and extraction of minerals in solution (U.S. EPA, 1973). Residual brines are disposed of through injection wells. Another operation practiced where salt deposits exist includes the construction of caverns for storage of liquid petroleum gas. In this procedure, water is injected into salt beds and caverns develop as salt is dissolved and brine material is pumped out. The extracted brines are then disposed of by injection wells.

The in-situ mining of metals, particularly copper, by injecting acid through wells into an ore body or tailings pile also causes elements to go into solution. Extraction of solutions containing metals via use of pumping wells or seepage is then accomplished (U.S. EPA, 1973). Oftentimes, deep injection wells are then used for disposal of spent acid solutions, once the valuable metals have been removed.

The problem of groundwater contamination from solution mining of soluble minerals and the techniques to prevent such incidents are very complex. Solution mining of metallic minerals presents many problems. This is because the mining will, in most cases, be in geologic strata containing usable waters. Therefore, the mining itself needs to be carefully managed to avoid groundwater contamination. Disposal of spent acid solutions by injection must be similar to other controlled industrial injections. McKinney (1973) and Pernichele (1973) discuss trends in solution mining and mining geohydrology.

Geothermal Wells

The Geothermal Steam Act of 1970 (Public Law 91-581) provided an impetus to development of geothermal energy. Condensed steam and cooled brines could be reinjected through wells into geothermal structures.

The Bureau of Reclamation anticipated at that time that production of 2.5 million acre-feet of freshwater per year could be produced from 3 to 4 million acre-feet of brines in the Imperial Valley of California. The desalted water that would be removed, however, had to be replaced with water from the Pacific Ocean or another source. This quantity of water would have to be injected through many wells along the periphery of the geothermal field. The same reservoir pressures would then be maintained and lowering of the overlying freshwater table would also be prevented (Bureau of Reclamation, 1972; from U.S. EPA, 1973). In addition, high pressures, temperatures, and corrosiveness of injected fluids would also cause problems in injection wells. Plugging of injection wells of course would also be difficult or impossible, if damage occurred to the subsurface casing.

Wells for Gas Storage

Underground gas storage may be defined as "storage in rock of synthetic gas or of natural gas not native to the location" (U.S. EPA, 1973). Storage of gas can occur in depleted oil or gas reservoirs, groundwater aquifers, mined caverns, or dissolved salt caverns. Gas storage occurs in gaseous or liquid form.

According to U.S. EPA (1973), quantities of gas are stored in the gaseous form in depleted oil or gas reservoirs or groundwater aquifers. In 1971 there were 333 underground gas storage fields in twenty-six states and 60 percent of the storage capacity was concentrated in Illinois, Pennsylvania, Michigan, Ohio, and West Virginia. The number of wells per field ranged from 10–100, depending on the size of the

initial structure in which the gas was being stored (American Gas Association, 1967, 1971).

Underground gas storage fields contaminate usable groundwater by upward leakage of material through cap rock, improperly abandoned plugged wells, improperly constructed and maintained gas injection, and/or withdrawal wells. Gas from overfilled fields also migrates laterally into storage aquifers which contain usable water. A leaky gas storage field in Illinois has been described by Hallden (1961). Gas storage fields are also subject to state or federal permits and regulations. The engineering characteristics of such fields must also be determined prior to issuance of permits. These fields must be monitored properly during their operation.

Release of contaminated or toxic liquids through wells into freshwater aquifers causes degradation of groundwater in the vicinity of gas storage facilities (U.S. EPA, 1973). Degradation then, can eventually spread over a wide region and extend into surface waters hydraulically connected with receiving aquifers. If water-level cones of depression around nearby operating water-supply wells are large enough to encompass injection sites, or if wells are downgradient along natural flow lines from injection sites, contamination of wells eventually takes place. Movement of the contaminated water from injection zones into overlying or underlying freshwater aquifers also occurs.

Cooling Water, Stormwater, and Domestic Sewage

Contaminated fluids intentionally injected through wells into freshwater aquifers, other than those from agricultural and mining wastes, include cooling water, stormwater, and domestic sewage.

When cooling water is returned to the same aquifer from which it was originally pumped, the chemical quality of the water is unchanged from its native state except for an increase in temperature. The increased solubility of aquifer materials due to this rise in temperature is insignificant, except in carbonate aquifers. Sequestering agents, such as complex polyphosphate-based chemicals added to the water to inhibit oxidation of iron, can also become a source of contamination.

Domestic sewage from individual households disposed of into household wells can be contaminated with numerous organic–inorganic substances (e.g., bacteria, viruses, biological organisms, solvents, etc.). This type of discharge receives no natural treatment during passage through the injection system, except for settling of the solids. The quality of municipal sewage effluent disposed of into injection wells also depends on the degree of pre- or final treatment before disposal and the initial source of the wastes (e.g., residential, commercial, industrial) occurs. Municipal sewage effluent consists mainly of domestic wastes with high dissolved solids and includes many nitrogen-cycle constituents, phosphates, sulfates, and chlorides. Municipal sewage in some localities also contains substantial amounts of industrial wastes. Different degrees of treatment remove and reduce certain constituents, but even with advanced treatment dissolved constituents such as heavy metals still remain. The problem of septage released into municipal sites can also be compounded by manholes used by septic tank and cesspool pumping companies, and regulations and tracking mechanisms for industrial and hazardous waste streams must be implemented and enforced.

The chemical quality of tertiary municipal sewage, native groundwater, and water recovered from observation wells, from an experimental injection study in Long Island, New York, showed that concentrations of ammonia, iron, phosphate, sulfate, other constituents and dissolved solids, were significantly high (Vechioli and Ku, 1972). Bacterial counts in the treated sewage were low, due to chlorination before injection.

Stormwater runoff can also have a low dissolved solids content. However, the initial slug of stormwater drainage may be contaminated with animal excrement, traces of pesticides, fertilizer nitrate from lawns, organics from petroleum products, rubber from tires, bacteria, viruses, heavy metals, organics, and other significant contaminants. If deicing salts are applied to roads in winter, the chloride content of stormwater runoff may also rise to several thousand mg/l.

Movement of Contaminants

The many natural processes that affect chemical transport from point to point in subsurface areas can be divided into three categories: physical, chemical, and biological (Keely et al., 1986). Contaminant transport in the subsurface is an undivided phenomenon composed of these processes and interactions. A collection of scientific laws and empirically derived relationships comprises the overall transport process. Additional efforts should be devoted to site-specific characteristics of natural process parameters, rather than relying exclusively on chemical analyses of groundwater samples.

Any well that produces water will also accept it. The rate a well accepts water is dependent on the nature of the injected fluid, hydraulic properties of the aquifer, amount of water removed, and related factors. In wells that penetrate permeable aquifers, water can be introduced via gravity methods at rates of several hundred gallons per minute or more. However, a well penetrating an inadequate aquifer accepts only small amounts of injection material, and if the fluid is injected under pressure, the rate of injection increases.

The rate of injection is also dependent on the permeability and thickness of aquifer materials, depth to natural water levels in the well, diameter of well, area and number of openings in the well screen, and chemical compatibility of injected fluids with native groundwater. If fluids being injected contain suspended material or air bubbles, rapid clogging of the aquifer occurs and injection rates fall off. Growth of bacteria and formation of chemical precipitates within the well and aquifer can also interfere with injection rates. Another further limitation on the rate of injection can also include natural or induced rises in groundwater levels. These rises in groundwater levels can cause an overflow of liquid materials onto the land surface.

Fluids injected through wells can also create local groundwater mounds in unconfined aquifers and pressure mounds in confined aquifers. These mounds can be described as mirror images of water-level cones of depression that develop around pumping wells. The configuration of specific mounds is usually symmetrical, with maximum rises in water level occurring at each well boring.

After injected fluid has entered the saturated zone, it moves radially away from the well. As this movement occurs, it displaces natural groundwater and creates a zone of mixed water along its perimeter. Contaminated water then moves towards the hydraulic gradient to a point of discharge. This point of discharge may be a well, spring, or a surface water body. If, however, injection wells initially leak due to corrosion of casing, casing breaks, or poor construction of the well, contaminated water then can move into freshwater aquifers above or below initial injection zones.

Wastes in Wells and Their Control

Where injection of wastes through wells into freshwater aquifers is proposed or is in progress, hydrogeologic investigations must be undertaken to prevent or control groundwater contamination. Hydrologic investigations should include:

- an evaluation of the groundwater environment and factors affecting its flow
- description of existing or planned nearby wells
- analyses of the directions and rate of movement of injected fluids, i.e., modelling should occur so projections can be made on how much time will elapse before contaminated water arrives at nearby wells
- studies to determine inter- and intra-aquifer movement of the injected water
- investigations concerned with the chemical, biological, and physical properties of injected fluids; the degree of pre-treatment needed; and the compatibility of injected wastes with the native groundwater
- evaluations on suitable locations and spacings of injection wells, including rate of injection
- considerations for future land use of the established and then abandoned injection well sites

Steps should also be taken to block the underground flow of the waste fluids or to remove fluids by pumping where the threat from contaminated groundwater is severe. Blocking of contaminants in some instances can also be accomplished by constructing subsurface barriers. Diversion flows by creating hydraulic barriers is another approach that can be implemented. Flow can be diverted by injecting water into wells installed across the known path of flow or by pumping from these same wells to induce contaminants to flow toward these barriers. Eventually, however, since pumping contaminated fluids back out of the ground creates new problems, facilities will eventually have to be then fitted for proper treatment and disposal of the later pumped water.

Alternatives and options to disposing of wastes through injection wells must be critically examined. Harmful effects must be avoided. Sewering for example, which exports waste materials, can have adverse effects because of loss of recharge and lowering of water levels. In the case of cooling water being returned to the aquifer from which it is drawn, an alternative is to use atmospheric heat exchangers instead of withdrawing water from the aquifer.

Halting the disposal of wastes into wells is desirable, but it should be recognized that stopping injection represents a partial contamination control measure. Fluids already injected will continue to contaminate the aquifer.

After the hydrogeologic environment has been studied and the mechanisms of contamination for an area defined, a systematic monitoring program should be implemented. This program would provide continuing surveillance of contaminated water, evaluate the efficiency of any control measures that may be instituted, or provide an early warning system for areas not yet contaminated. Depending on local conditions, a series of properly constructed monitoring wells at different depths in the known contaminated areas, clean areas, and at scattered locations can be constructed. Periodic and systematic monitoring of these wells for chemical content of groundwater and changes in groundwater levels provides valuable data on the behavior of underground contamination and on potential environmental threats to water supply wells.

Use of Lagoons, Basins, and Pits to Dispose of Liquid Wastes

Unlike septic tank and leachfield disposal systems or landfill operations, lagoons, basins, and pits are usually open to the atmosphere, although some pits and small basins may be placed under rooftops (U.S. EPA, 1973). Some of the lagoons, basins, and pits are also designed to discharge liquid into soil systems and groundwater areas, whereas others are constructed of watertight materials. Unlined and lined structures can therefore be used, but each by its design or failure, becomes an important concern in terms of groundwater quality. Certain regulations may also limit the amount of discharge.

Lagoons and basins can be adapted to many municipal and industrial uses including storage, processing, or waste treatment of large amounts of liquids or sludges. Lagoons or basins can also serve as large septic tanks for raw sewage, secondary or tertiary sewage oxidation ponds, or as spreading basins for effluent from wastewater treatment plants by groundwater recharge, provided effluent regulatory standards are met. In industry, some systems can serve as cooling ponds or for holding hot wastewater until it cools down. Wastewater can also be stored for later treatment and discharge. Some of these wastes, depending on effluent quality can also be applied to land during various times of the year. Lagoons are even used as evaporating ponds to concentrate and recover salt or other chemicals.

Basins that are lined are used as evaporation ponds to concentrate salts, brines, minerals, etc. Recovery of precious minerals, regulatory concerns, or economic disposal of concentrates often are the motivating factor. Basins are also used in oil fields, refineries, and chemical processing plants as holding sumps for brines or wastes. Often, these wastes are held for later disposal by deep-well injection or other means. Lined basins have also served as a sump receiving wastes from fruit and vegetable canneries. Lined basins are also used to dispose or hold stormwater from roof drains and highway underpasses.

Lined pits have historically been used by industry and municipal sewerage works for processes ranging from tanning of animal hides to storage of metal plating wastes. Although lined pits are constructed of concrete, metal, or fiberglass, undetected leakage of highly concentrated contaminants, if it occurs, can have a significant detrimental effect on groundwater quality.

Hazards to Groundwater

Sewage lagoons can degrade groundwater quality to the same degree that septic and leachfield systems do. McGauhey and Krone (1967) described how a continuously inundated soil becomes clogged so that the infiltration rate of fluids is reduced below the minimum for it to be considered an acceptable infiltration system. In addition, if the groundwater surface is close to the lagoon bottom, a hanging column of water will be supported by surface tension and the soil will not drain properly. Clogging then continues indefinitely, even if new liquid is not added. Ponds must therefore be designed to be loaded and rested intermittently. This is necessary to maintain acceptable recharge rates (McGauhey and Krone, 1967). If, however, isolating liquid concentrations from the lagoon and groundwater is important, then a low infiltration rate will still cause an undesirable quantity of contaminated water to pass through the water–soil interface. Consequently, the contaminants continue to be carried downward with percolating water to the groundwater.

Lagoons and pits used to recharge groundwater directly with runoff from highways and roofs can also affect groundwater adversely, especially in areas that have high permeable soils, and distances to the water table are low. Oils and heavy metal accumulation from road surfaces are added to rainwater and deposited in pits. Lead has been described to be significantly higher on soils adjacent to highways. Its movement to groundwater also occurs.

Liquids percolating from lagoons or basins that originated in some industrial operations can degrade groundwater to a higher degree than domestic sewage. Chromates, gasoline, phenols, picric acid, and miscellaneous chemicals are known to have travelled large distances with percolating groundwater (Davids and Lieber, 1951; Harmon, 1941; Lang and Gruns, 1940; McGauhey and Krone, 1967; Sayre and Stringfield, 1948). Liquids stored in lagoons, basins, and pits may also contain brines, arsenic compounds, heavy metals, acids, gasoline products, phenols, radioactive substances, and other miscellaneous chemicals, again dependent on their source of origin.

However, where storage areas holding these wastes have been actively used for years, leakage eventually seems to occur through the sides and bottom of some of the particular lagoons or basin areas. Their cumulative effects on groundwater can be significant, and any plume of contaminated liquid may have traveled long distances by that time. The first realization that extensive groundwater contamination has occurred comes when the plume reaches a natural discharge area at a stream, and effects on surface waters are noted, or pumpage of the contaminated water shows up in a water supply well.

A basic example of environmental consequences and a leaky basin containing metal-plating waste effluent from industrial plants is given in Perlmutter and Lieber (1970). Plating wastes containing cadmium and chromium seeped from disposal basins into the upper glacial aquifer of southeastern Nassau County, New York. The seepage formed a plume of contaminated water that was identified to be over 4,000 feet in length, 1,000 feet wide, and 70 feet thick. At the end of the plume contaminated groundwater was being discharged naturally into a small creek that drained the aquifer. The maximum observed concentration of chromium in the groundwater was 40 mg/l, and levels of cadmium were observed as high as 10 mg/l.

Control Methods to Protect Groundwater

When lagoons or basins are used for deliberate disposal of sewage effluents or surface runoff by groundwater recharge, controls pertinent to groundwater protection are essentially self-generating. The disposal system will not work if designed improperly. Moreover, engineering and hydrogeologic knowledge should prohibit the construction of systems directly in the groundwater. Adequate distances should also be allowed between the infiltrative surface and the groundwater to permit drainage. Construction should also be prohibited in faulted strata and unsuitable soils. An important control measure in groundwater protection from use of lagoons or basins is, therefore, to apply state of the art knowledge to their siting and design, build them properly, and not overload the system. Local regulations should also be followed.

Control of industrial waste discharges to groundwater is complex. In states with highly organized water pollution control agencies (e.g., California), permits are issued on the basis of design and surveillance programs being put into place. Because of the variety and complexities of industrial wastes and of the different situations in which they occur, control of groundwater contamination by such wastes is dependent upon proper design of new systems and modifications and correction of malfunctioning existing systems. Methods for controlling groundwater contamination from industrial lagoons, basins, and pits have been summarized by U.S. EPA (1973), and include:

- pre-treatment of wastes for removal of toxic chemicals
- requiring that lagoons, basins, and pits containing toxic fluids be lined with impervious barriers
- barrier wells to pump and intercept plumes of contaminated groundwater from existing industrial basins where leakage has occurred
- no brine pits, unless they are inspected and control measures can be enforced, if necessary
- that the location and identification of unauthorized pits on industrial sites be evaluated and designed on a case by case basis, and appropriate regulatory action be applied to these routinely

Lagoons, basins, and pits can represent multiple point sources of contaminating factors significant to groundwater quality. A program involving monitoring wells is critical and necessary. Periodic sampling and evaluation of data from existing and newly constructed wells selected for their potential to reveal groundwater quality and point contamination is one approach. Monitoring of control measures themselves is also necessary, to assure that groundwater protection is being accomplished.

NINE

Underground Tank and Pipeline Leakage

In the United States underground storage and transmission of fuels and chemicals is a common occurrence within commercial, industrial, and individual uses (American Petroleum Institute, 1983a, 1983b). Pipes and tanks, however, are subject to structural failures arising from a wide variety of causes. Subsequent leakage then becomes a source of contamination that can affect the quality of local groundwaters.

Leakage of petroleum and its products from underground pipelines and tanks occurs more often than generally realized. This is particularly true for small installation areas, such as the use of oil tanks in the home and privately owned gasoline stations, where installation, inspection, and maintenance standards are low or nonexistent. In Maryland, in 1971, where standard investigative procedures were in effect at that time, sixty instances of groundwater contamination were reported in a single year from gasoline stations (Matis, 1971). Such instances have increased. In northern Europe, in 1973, where homes are heated by oil stored in subsurface tanks, oil contamination was also a major threat to groundwater quality (Todd, 1973; from U.S. EPA, 1973). The problem is currently widespread and a major concern and priority in the United States.

Historical Aspects

Traditionally, pipelines and storage tanks were buried because the cost of burial usually did not exceed the potential damage to them by weather, vandalism, etc. The aesthetic value of burying utilities and other commercial or industrial structures undoubtedly at one time increased the number of tanks and pipes that were placed in excavations. However, because pipelines are an economical means of shipping, there always has been an increasing trend toward developing and improving methods for movement of solids in pipelines, such as coal or ores by powdering and mixing them with water or oil to produce pumpable slurries (U.S. EPA, 1973).

A significant problem associated with underground storage tanks (USTs) is that their installation and use are not usually regulated. If regulations exist concerning such tanks, they usually are local in origin and require only that tank construction and installation be satisfactory. It is rare for routine or mandatory follow-ups or periodic checks to occur. Since many of these tanks are small, cathodic protection is not always required, even when the tanks are located in clay soils, which are known to promote galvanic action.

Pipelines are used for transportation, collection, and for distribution of liquids or slurries (U.S. EPA, 1973). Pipelines transport many materials, including oil, gas, ammonia, coal, and sulfur. Their heaviest use, however, involves petroleum products, natural gas, and water. Accidents summarized for 1971, in order of magnitude of occurrence, included: crude oil, gasoline, propane gas, fuel oil, diesel fuel, condensate, jet fuel, natural gasoline, anhydrous ammonia, kerosene, and alkylate (Office of Pipeline Safety, 1972).

In addition, many industries employ underground collection pipeline devices to move process fluids and wastes within a plant or to move materials to storage areas or ship them out. Collection pipelines within oil field areas are used to move crude oil from wells to tanks where separation of brines, storage, and shipment occurs. Because interstate pipelines are also a major means of transporting materials, they are regulated by the federal government (Department of Transportation). These regulated pipelines are required to report leaks and spills because leakage of petroleum products produce fire or explosion hazards. Generally, leakage associated with local distribution systems, gas stations, residential storages, and even relatively large intrastate carriers are not always reported. Leakage that is reported, however, is probably small compared to the total amount for the entire country.

Environmental Consequences

Pipeline and tank leakages of petroleum and fuels into soils impact several environmental concerns, depending upon the specific product which has leaked, and its amount. Oils and petroleum products in trace quantities also render private or public water supply systems objectionable (i.e., because of taste, odor). However, if concentrations of petroleum products, liquified gas, and natural gas are high, then vapors eventually seep into basements, excavations, tunnels, and other underground structures. These vapors then mix with air and constitute a severe explosion or fire hazard if exposed to open flames, sparks, or ignition of some source.

Chemicals such as ammonia and other agricultural or industrial products also have toxic properties and will add to the nitrification of groundwater, change the pH, and accelerate the solubility of soil solids with heavy metals.

Factors Causing Leakage

Corrosion is a major cause of leakage in systems. It attacks supply lines externally and internally (U.S. EPA, 1973). Another major cause of leakage occurs as a combined result of pipeline components, equipment, personnel failure, and malfunction of systems. Rupturing of lines as the result of construction equipment working in the vicinity of systems also causes leakage. Vandalism (e.g., bullet holes in exposed sections of pipe tanks or valves), cold weather, lightning, floods, and earthquakes also can cause leakage. Other related causes of leakage include: forest fires, defective pipe seams, leaking of gaskets, improper repair welds, malfunctioning of valves, malfunctioning of control or relief equipment, defective girth weld, threads stripped or broken, and pump packing failure.

Pipelines therefore should be routed to avoid areas with a history of such acts of nature and structurally engineered to withstand natural calamities.

Movement of Contaminants from Underground Areas

Leaks into underground excavations behave in several ways, depending on soil characteristics and the depth below the leak of the saturated zone (U.S. EPA, 1973). The statements that follow apply not only to oil but also to other liquid contaminants released from underground storage tanks and pipelines.

If the leak is from a tank of limited horizontal extent or from a pipeline in permeable soils, the liquid remains in the vicinity of the leak and moves downward through the soil under the influence of gravity. If the leak occurring comes from a pipeline constructed in impermeable soils, the resultant liquid remains in the trench. In a sloping trench situated within impermeable soil, the leaked fluid will move through backfill material trenches along the outside of the

pipe towards the slope and as the leaked liquid penetrates the soil under gravity, it also coats soil particles. This process of coating soil particles removes fluid from the downward moving body and if the quantity of leaked liquid is minute, it stops. However, this leaked liquid will not remain immobilized because subsequent rainfall can wash the contaminant from soil particles and carry it downward to the saturated zone.

If the leakage volume that reaches the saturated zone is high, its path of movement depends upon density and viscosity of the initial liquid and its miscibility with water. Miscible liquids in this case mix and dilute slowly, depending on distance and time involved. Subsequent precipitation also displaces oil or other low density fluids lying on the upper portions of the water table and thereby produces a mixture extending into the saturated zone. Once the initial leakage is in the saturated zone, a contaminant can move down gradient in the direction of groundwater flow.

Leaked liquids also move laterally for great distances above saturated zone areas. If a spill is large or if a leak continues over a long period of time, the fluid migrates along impermeable layers above the water table. A similar situation can occur along a pipeline. Liquid will move along the pipeline, until a permeable soil is reached and here it penetrates downward. Once this happens, however, it should be recognized that chemicals do not always move at the same velocity as groundwater (U.S. EPA, 1973). Sorption, varying miscibility, and solubility rates with water (including different chemical activity rates and reactions that occur with soils) usually move through various soil types and horizons in the direction of the groundwater flow, but do so more slowly (Comm. on Environmental Affairs, 1972).

Control Methods

Methods for controlling and abating contamination of groundwater by leakage from tanks and pipelines in underground excavations have been concerned primarily with reducing fire, explosion, and toxicity hazards. Although reduction of fire and explosive hazards is not aimed at abating contamination of groundwater, it may be. This is because methods for handling hazardous materials also apply to handling leakage materials, e.g., sewage, brines, agricultural and industrial chemicals. While the EPA has proposed new guidelines in 1987 which are discussed in the next section, control methods to *prevent* leakage should emphasize prevention activities such as:

- application of satisfactory corrosion preventing coatings, such as tar or plastic used on the outside of tanks and pipelines
- use of cathodic protection to minimize corrosion resulting from galvanic action (Department of Transportation, 1969)
- use of internal fiberglass linings, which will not

deteriorate, for small tanks such as those used for gasoline storage (Matis, 1971)

Storage sites for location of tanks and pipelines can also be designed to contain liquids that have leaked. In many instances, fluids can be trapped and removed before they penetrate soils. Although these methods pertain almost exclusively to tanks, some modifications can be made to pipeline areas or portions thereof. Excavations used to enclose subsurface tanks can be lined with impermeable materials such as clay, tar, sealed concrete, or plastic linings. Double or triple linings can also be installed in many instances.

Within pipeline areas, containment can also be accomplished by use of automatic shut-off valves inserted in the pipe at selected intervals. These valves can be designed to close off different sections of pipe, specifically, when line pressure drops occur. This method when put into practice limits the spread and volume of leaks. In 1973, this form of protection was required on interstate pipelines but not on small collection and distribution systems (U.S. EPA, 1973).

Once a leak is discovered and can be accessed after it occurs, a method for preventing groundwater contamination at that time is to remove the soil that has been soaked with leakage fluids. This method must be applied before significant rainfall or precipitation occurs. This is because, without the flushing action of rainfall or precipitation, liquids eventually move downward, although at a slower pace, by gravitational pull. According to Todd (1973), however, several hours, days, or weeks may be available before these penetrating liquids reach depths beyond those considered reasonable for earth removal. This migration downward seems to be a characteristic of small leaks and of large leaks that eventually reach saturated zone areas. In dealing with large leaks that are associated with catastrophic failures, such as a big tank or pipe rupture, it is important to initiate cleanup procedures immediately.

The problem of how and where to dispose of soil also must be addressed. The most suitable method for handling biodegradable materials is to spread the contaminated soil in thin layers and permit the natural aerobic soil bacteria to degrade it. If contaminated soils are not biodegradable, they must be removed and transported to appropriate hazardous waste disposal site areas. Earth removal is an extensive operation and requires much more than digging a hole and hauling soil off-site. However, this also might not be the solution. Certain local regulations relating to release of toxics into the air may prevent this.

In cases where contaminants have reached the water table but not moved significant distances from the initial leakage site, removal wells can be installed. If the contaminant has moved so far down gradient that recapture by use of drawdown cones established by newly drilled wells is not feasible, a ditch placed across the plume of contamination in some instances can be used to capture or contain the unsatisfactory material.

In addition, when the water table is at a great depth below the ground, many wells may be installed. These wells would be located across the plume of contamination in order for the effects of their combined drawdown cones to produce a trench in the water table. The contaminant then should not be able to escape from the artificially created depression. Through time, therefore, the contaminants will be removed by the wells provided no new contaminated fluids enter into the same aquifer based system. Also, during this pumping process as contaminated water is removed, it must be treated as an industrial wastewater effluent. One should also assume that various contaminants may be present, so disposal to a municipal sewage system or return to groundwater concerns may not be feasible. The appropriate methods and techniques used to pump water and return it for each specific occurrence depend on the nature of original contaminants, available wastewater treatment facilities for the area, and regulations concerning the disposal of such wastes.

Monitoring Procedures

Monitoring required to detect leaks from tanks and pipelines in excavations should be in proportion to the quantity of chemicals handled and the risks involved if any leakages occur. Household fuel storage tanks, gasoline station storage tanks, and local collection or distribution systems, must be monitored in some way for leakage. Local governments in turn, must regulate such operations by ordinances and codes specifying the materials, methods of installation, and acceptable and verified monitoring methods and techniques. Many leaks will occur from larger underground pipes and tanks as well as from these small systems, and monitoring for such leaks should not be by "discovery" (i.e., when a nearby owner of a water well discovers that the well is contaminated). Owners or operators of systems should not discover that their chemicals or fuels are disappearing from their storage or transport facilities during high groundwater table rises that may occur above the level of a leak in a tank so that the owner or operator discovers that the tank during that period supplies water rather than chemicals or fuel.

Monitoring methods and procedures for interstate carriers for pipeline safety are under the control of the Office of Pipeline Safety (OPS). To assure that interstate carrier monitoring and pipeline operation are being done satisfactorily, the OPS has required reports of leakages from initiation of the leak to the time it is corrected (Department of Transportation, 1969; Office of the Secretary of Transportation, 1970).

In addition, because of OPS, pipelines contain pressure-monitoring devices that automatically close valves to isolate a section of pipe whenever a significant pressure loss occurs (Department of Transportation, 1969, 1970). Systematic evaluation of pipelines and tanks are also accomplished by consistent monitoring, periodic inspection, and routine pressure testing (Dept. of Transportation, 1970a; Office of the Secretary of Transportation, 1971, 1972).

Throughput monitoring, another method of monitoring potential leaks, compares input and output of volumes. This method of monitoring detects large leakage rates, but small volumes go undetected. However, improved instrumentation permits the detection of small leaks, but usually they are only pinpointed by routine inspections and pressure testing.

Periodic inspections for potential leaks should also include an on-site survey and detailed visual evaluations. If volatile chemicals are involved, pipes can be inserted into soil areas and air samples examined through gas detectors. The periodic evaluation of pipelines can be conducted by individuals on foot, vehicle, or aircraft. In most cases the dominant method of detection involves visual means (i.e., by observing the effects that leaking chemicals have on vegetation).

Tanks in lined excavations can also be examined by use of liquid level or vapor sensors placed between the linings in excavations and the tanks. These in turn connect to alarm systems located where personnel observe sensor readings. Pressure tests can also be conducted on pipelines and tanks after repairs and in situations when corrosion is detected. A tank or section of pipeline consequently is filled and pressurized with results monitored. Allowances and adjustments, in this instance can be made for temperature changes, expansion of materials under pressure, and degrees of tightness. Normal pressure-testing then occurs for a twenty-four-hour period, and reports are filed with operators and the Office of Pipeline Safety. Pressure testing if done properly is more sensitive than throughput monitoring and periodic evaluations or surveys. However, persistent small leaks can still go undetected, since some of the tests conducted to determine leakage are equivocal, potentially inconclusive, and not 100 percent reliable.

New Proposed U.S. EPA Regulations on Underground Storage Tanks

The U.S. EPA has recently proposed regulations for many of the nation's underground storage tanks (USTs) (U.S. EPA, 1987). The complete proposal appears in the *Federal Register* (April 17, 1987). The proposed technical requirements include the steps the tank owner or operator may have to take to help protect drinking water resources. These steps are intended to avoid the high cost of cleaning up the environment and settling legal suits that can result if tanks leak. Four major points in the proposed regulations include:

- New USTs must meet the requirements of the Interim Prohibition Standards.
- Within three to five years after the regulations

become final, USTs will have to meet leak detection requirements.

- Within ten years, USTs must also be protected from corrosion and equipped with devices to prevent spills or overfills.
- Owners will be financially responsible for the cost of cleaning up a leak and compensating other people for injury and property damage caused by leaking tanks.

A UST is any tank, including any underground piping connected to the tank, that has at least 10 percent of its volume below ground (U.S. EPA, 1987). The proposed regulations will apply only to USTs storing either petroleum or certain hazardous chemicals.

The kinds of tanks not covered by these regulations include:

- farm and residential tanks holding less than 1100 gallons of motor fuel used for non-commercial purposes
- tanks storing heating oil burned on the premises where it is stored
- tanks on or above the floor of underground areas, such as basements or tunnels
- septic tanks and systems for collecting storm water and wastewater
- flow-through process tanks

In addition to leaks from tanks and piping, the UST can be the site of accidental spills and/or overfills. These spills are usually the result of human error and can be avoided by following correct tank filling practices. Mechanical devices, such as overfill alarms, also reduce spills and overfills. Many leaks also result from piping failure. Piping fails more often than tanks because it is smaller and less sturdy. It may suffer from installation mistakes and the effects of corrosion. The proposed regulations will apply to the entire UST system—both tanks and piping.

The goal of the proposed federal regulations over the next ten years is to improve petroleum USTs already in the ground so that they will meet more demanding requirements. All USTs at the end of ten years in the ground, now will need to show three required improvements that include:

(1) Meeting the same requirements for corrosion protection that apply to new USTs

(2) Meeting the new UST requirements for having a leak detection method

(3) Being equipped with devices that prevent spills and overfills

The proposed UST regulations apply to the same hazardous chemicals identified by CERCLA [Section 101(14) of the Comprehensive Environmental Response, Compensation, and Liability Act of 1980, or Superfund]. If a UST contains both petroleum and hazardous chemicals, it will be regulated as a chemical UST if the principal part of the stored mixture is made up of hazardous chemicals.

In addition to the proposed requirements for new petroleum USTs described earlier, chemical USTs would need to satisfy "secondary containment" and "interstitial" monitoring requirements. All new chemical USTs must have secondary containment. The UST itself makes up the first or primary containment. Using only primary containment, a leak can escape into the environment. By enclosing a UST within a second wall, leaks can be contained in a relatively small and controllable area.

Several ways can be used to construct secondary containment. These include (U.S. EPA, 1987):

- placing one tank inside another tank (making them double-walled systems)
- placing the tank inside a concrete vault
- lining the excavation area surrounding the tank with natural or synthetic liners that cannot be penetrated by the chemical

The chemical UST must have a leak detection system for interstitial monitoring that can indicate the presence of a leak in the confined area between the primary and secondary walls. Several devices are available to monitor this confined "interstitial" area. The proposed regulations describe these various methods and the requirements for their proper use (U.S. EPA, 1987).

USTs must have a "leak detection method" that provides monitoring results at least every thirty days. Leak detection methods must be capable of effectively detecting the stored product. They must be able to detect a leak in any part of the UST, including the piping. Leak detection can consist of one or a combination of the following methods (U.S. EPA, 1987):

(1) Tank tightness testing and inventory control
(2) Automatic monitoring of product level and inventory control
(3) Monitoring for vapors in soil
(4) Monitoring for liquids in the groundwater
(5) Monitoring for interception barriers
(6) Interstitial monitoring within secondary containment
(7) Other methods approved by EPA (many other methods are still being developed)

ABANDONED TANKS

Regulatory requirements for abandoning USTs vary from state to state—even county to county—and it is important for facilities to be sure their USTs are abandoned properly (Robbins, 1987). A sound tank abandonment program should be divided into three stages:

(1) Initial planning
(2) Plan implementation
(3) Site closure and documentation

Groundwater Contamination by Surface Waters, Atmospheric Precipitation, Salt, and Salt Intrusion

Scope of the Problem

Groundwater contamination by flow from surface water is known to take place in some portions of the country (Fuhriman and Bargon, 1971). Undoubtedly, it is also occurring in many other places, yet undetected. Water applied during irrigation and that which temporarily covers surface areas during floods are examples of how the mechanisms for contaminants enter the soil, pass downward into aquifers, and become established. Surface waters from open bodies (e.g., rivers, lakes) can enter into aquifers where groundwater levels are lower than surface water levels. Pumping from shallow wells near a stream and infiltration of water into soil from flash runoff occurrences in dry stream channels are other methods of how this type of groundwater contamination can occur.

Surface runoff in urban areas also contains many contaminants with the infiltration and percolation of this water causing an increase in dissolved constituents in groundwater. The kinds of contaminants that enter aquifers through flow from surface to groundwater include the entire spectrum of inorganic and organic compounds as well as bacteria, viruses, and other biologicals.

The reverse situation, that of groundwater contaminating surface water has received little attention. However, flowing groundwater will discharge into surface water bodies if the water table is higher or if the surface water is lowered by pumping wells. A large fraction of streamflow is also derived from drainage of groundwater. Long-term degradation of surface water can therefore be anticipated in areas where groundwater contamination continues to exist and a definite interplay with the surface water occurs. In addition, in some instances the flow of groundwater into a lake can also slow the rate of lake acidification (Anderson and Bowser, 1986).

Environmental Consequences

Contaminants entering groundwater from surface-water sources gradually disperse and ultimately affect the quality of the volume of groundwater if the source continues. Dissolved solids usually move farthest with groundwater and form a plume that extends thousands of feet downstream from a point source of contamination (U.S. EPA, 1973).

Nature of Contaminants

Stormwater runoff can exhibit a wide range of inorganic, organic, bacteriological, other biologicals, and chloride contents. Nitrates, lead, chromium, other heavy metals, pesticides, and organic compounds have been reported in stormwater runoff, specifically from urban areas (Sartor and Boyd, 1972; from U.S. EPA, 1973). Road salting in parts of the United States during winter months can also be a substantial contributor of chloride to groundwater (Hanes et al., 1970).

Water in streams consists of base flow fractions from seepage of groundwater and overland runoff fractions that are usually more mineralized than the initial groundwater. Other factors that also affect the quality of surface waters include: geochemical reactions which occur between the liquid interface, streambed, and suspended materials; evapotranspiration; and the activity and degree of biota in the stream. In addition, superimposed on these natural processes are contamination from human activities (anthropogenic sources).

The chemical quality of streamflow ranges widely, and in many large undeveloped drainage basins in upstate New York and in New England the water is very soft and has a dissolved solids content of less than 30 mg/l (U.S. EPA, 1973). In contrast, the Hudson River in southern New York receives large discharges of sewage effluent, including industrial and agricultural wastes. Effluents from chemical, paper manufacturing, fruit and vegetable processing, canneries, and other sources also empty into this drainage system. Large amounts of warm water are also returned to streams after use in cooling systems by some utilities and industrial concerns. All of the previously mentioned sources affect groundwater contamination. Nonetheless, the quality of water pumped from wells near surface-water sources generally reflects an integration of native groundwater and surface water that has infiltrated during the different seasons.

Lake water and its composition are affected by many of the same factors influencing streams, including incomplete mixing of water within a lake, thermal stratification, evaporation, and the character of subsurface, surface inflows, and in some cases direct precipitation into a lake. Surface waters also have higher annual temperature ranges when compared to groundwater (Rorabaugh, 1956). Consequently, induced infiltration from surface sources can cause temperature changes from 5–10°C in groundwater. This lag between changes in temperature between a surface-water body and the interconnected groundwater area also reflect summer or seasonal infiltration, while below-normal temperatures usually reflect winter infiltration of surface water, in areas where extreme temperatures occur between summer and winter.

Movement of Contamination

In places where surface-water bodies are closely connected to underlying groundwaters, water moves in the direction of the hydraulic gradient (U.S. EPA, 1973). In different reaches of a stream channel, the water flow gains or loses volume, i.e., receives groundwater inflow if the water table in the area is high, or loses water to an aquifer if the water table is low. Flow from surface areas to groundwater also takes place where wells are installed near streams or lakes and the water pumped from the aquifer is replaced by induced infiltration of surface water. Methods for calculating stream depletion quantities by pumping wells have been discussed by Jenkins (1970). Field tests on calculating these quantities were also conducted at Kalamazoo, Michigan, by Rorabaugh (1956). The previous investigators also observed that under long-term steady flow conditions, most of the water pumped from a well comes from nearby interconnected surface-water sources. Basically, what appears to happen is that the quantity of surface water that moves into an aquifer depends on the transmissivity of the aquifer, the hydraulic conductivity, the area of concentration at the bottom of streams or ponds, the pumping rate, and the amount of surface water available.

Effective programs to improve the quality of water in streams, ponds, and lakes, through control of waste disposal and storm runoff are also effective in removing contaminant sources to groundwater initially fed to these sources by surface-water bodies. Control of pumping quantities and proper siting of wells through aquifer testing and other hydrogeologic evaluations can help to minimize seepage of contaminated surface water into aquifers.

Monitoring Procedures for Surface Waters

Monitoring the quality of surface-water bodies where groundwater can be affected provides data needed to define areas where surface water contaminates aquifers. Periodic water sampling and chemical analyses of water from existing or recently installed wells should be conducted in areas where the threat of contamination occurs from nearby surface-water bodies. Chemical analyses in instances such as these should be tailored to detect contaminants involved. Special observation wells may also be warranted in these areas to provide advance warning of significant differences in flow between surface and groundwater areas.

Groundwater Contamination from Atmospheric Precipitation

Raindrops falling through the atmosphere pick up different concentrations of dissolved solids from the particles being carried. Some of these particulates being carried originate from natural sources, while others are a result of human activities. High concentrations of suspended particulates result from human activities polluting the air, specifically, from industrial, automotive, and urban sources. Where air pollution levels are high, raindrops dissolve enough solids that eventually reach the earth's surface and become an initial source of groundwater contamination.

Contaminant Definition

Precipitation, including rain, snow, sleet, and dry fallout of particulate matter, differs from place to place, and through different time intervals. The great diversity in composition of the precipitation material has been discussed by Carroll (1962) and Gambell and Fisher (1966). Dissolved and particulate matter in precipitation material can be local or transported from other areas, including the oceans, different land masses, different vegetation habitats, industrial processes, variable fertilization and cultivation practices of the soil in agricultural areas, combustion of fuels, and other related sources. The chemical makeup of precipitated material collected in the United States prior to 1962, included recognition of sodium, potassium, calcium, chloride, sulfate, nitrate, and ammonia (U.S. EPA, 1973). The concentration of the previously listed constituents in this study were below recommended standards in drinking water, but even at these low concentrations, the annual fallout can be extrapolated to represent several tons per constituent per square mile. Other constituents analyzed in precipitation samples include iodine, bromine, iron, lead, cadmium, and traces of other heavy metals (Biggs et al., 1972; from U.S. EPA, 1973). Also identified in precipitation samples are silica, detergents, nitric and sulfuric acids, radioactive substances, and other chemicals. Precipitation is also acidic for some areas, shows a pH averaging 5–6, and is classified as acid rain.

Concentrations of heavy metals, organics, and radioactive substances as particulate matter in the air also include: benzene, soluble organics, nitrates, sulfates, antimony, bismuth, cadmium, chromium, cobalt, copper, iron, lead, manganese, molybdenum, nickel, tin, titanium, vanadium, zinc, and many complex compounds (Federal Water Pollution Control Agency, 1969).

Movement of Contaminants

That portion of precipitated water that infiltrates into ground areas and is not lost by evapotranspiration can be expected to contribute significant volumes to aquifers. Minerals that settle in the soil when rainfall evaporates will also be carried down to groundwater levels by subsequent infiltration of precipitation.

The portion of precipitation that reaches the groundwater varies from zero in arid regions to a high percentage in areas receiving moderate–heavy moisture on highly permeable soils. Contaminants present in rainwater and those picked up as water passes through soil particles, become active contaminants when they and the water that carries them reach the groundwater table. Contaminants present in precipitation falling on surfacewater bodies may also later reach groundwater levels at a downstream location or show up in various springs or lakes.

Control Methods

In order to control contaminants in precipitation, improvements in air quality must first be accomplished. These improvements can be brought about through regulations and enforcement actions that lead to a reduction of specific contaminants in the air. Emission standards from stationary and non-stationary sources must also be put into compliance, so they do not contaminate the atmosphere or produce extreme amounts of particulate polluted matter.

Saltwater Intrusion

Seawater in Coastal Aquifers

The problem of saltwater intrusion and upconing to a pumping well has been widely recognized in groundwater utilization for many aquifers (Huyakorn et al., 1987). In the United States, intrusion of saline water has resulted in the degradation of freshwater aquifers in at least 20 of the coastal states (Newport, 1977). Both vertical and lateral intrusions occur and are usually caused by overpumping in sensitive portions of a water supply aquifer.

Fresh groundwater in coastal aquifers is discharged into the ocean at or seaward of the coastline. If, however, groundwater demands become large, the seaward flow of groundwater is decreased or reversed. Seawater then advances inland within the aquifer and produces seawater intrusion.

The quantity of seawater intrusion has increased in coastal population centers, specifically, those that have overdeveloped their local groundwater resources to meet water supply needs. Early published reports, dating from the mid-nineteenth century in England, describe the increasing salinity of well waters which occurred in London and Liverpool at that time (U.S. EPA, 1973).

Most coastal areas of the United States have some aquifers contaminated by the intrusion of seawater (Task Committee on Salt Water Intrusion, 1969; Todd, 1960). Florida, California, Texas, New York, and Hawaii are among the more seriously affected.

In Florida the problem relates to permeable limestone aquifers, lengthy coastlines, and the desire of most inhabitants to live near coastal beaches, although seawater intrusion has been identified in many specific locations (U.S. EPA, 1973), and many municipal water supplies have been adversely affected since 1924. In the Miami area, intrusion has been a problem for some time, and in this region it was augmented by drainage canals which lowered water table areas and allowed seawater to advance inland (Parker et al., 1955).

In California, the large urban areas which concentrated high human population numbers in coastal zones also contributed to seawater intrusion in many localities (California Department of Water Resources, 1958; from U.S. EPA, 1973). Most of the coastal areas affected in this state contained confined aquifers, and the proportionate increases in salinity can be significantly correlated with lateral movement of seawater induced by overpumping of municipal wells.

In Texas, intrusion has been a problem in Galveston, Texas City, Houston, and Beaumont-Port Arthur (Pettit and Winslow, 1957). Saline water moves up-dip from the Gulf of Mexico through the confined Coastal Plain sediments. In New York, the problem is centered around the heavily pumped western portion of Long Island (Luscynski and Swarzenski, 1966). In Honolulu, the aquifers have been invaded by seawater due to overdraft conditions (Todd and Meyer, 1971; Visher and Mink, 1964).

Environmental Consequences

Because of its high salt content, 2 percent of seawater mixed with fresh groundwater makes the water unusable in terms of U.S. Public Health Service Drinking Water Standards (U.S. EPA, 1973). A small amount of intrusion therefore can threaten the continued use of an aquifer for water supply. Also, once invaded by seawater, an inland aquifer can remain contaminated for long periods. The normal movement of groundwater precludes any rapid displacement of seawater by freshwater. Abandonment of the underground resource may be a necessity, and treatment is oftentimes very expensive.

Causal Factors

The main contributor to seawater intrusion in coastal aquifers is overpumping. Pumping lowers groundwater levels and reduces freshwater flow to the ocean. This seems to occur even with an initial seaward gradient. During these same conditions, seawater still advances inland. If the pumping of freshwater aquifers reverses the gradient, the freshwater flow ceases, and seawater then moves into the entire aquifer.

In flat coastal areas, drainage channels also can cause saltwater intrusion. A reduction in water table evaluation occurs and it causes an associated decrease in underground freshwater flow. Tidal action can also carry seawater long distances inland through open channels. Here, it may infiltrate and form fingers of saline water that adjoin channels in the area.

In oceanic islands, freshwater aquifers form a lens overlying seawater. Consequently, if a well within the freshwater zone is substantially pumped, the underlying seawater will rise and contaminate the well. Wells in this instance can also serve as means of vertical access to freshwater aquifers lying above or below saline zones.

Contamination Movement of Saline Waters

A parabolic form exists in the interface between underground fresh and saline waters (Cooper et al., 1964). Saltwater in this area underrides the less-dense freshwater. When equilibrium is established, however, seawater then becomes stationary. Freshwater at this time then flows seaward. The length of the intruded wedge of seawater then varies inversely and proportionally in volume with freshwater flow. A reduction of freshwater flow, therefore, causes intrusion. However, flow reversal is not always required to achieve intrusion.

Seawater intrusion also involves miscible displacement of liquids in porous media, and diffusion and hydrodynamic dispersion mixes the two fluids (U.S. EPA, 1973). The ideal interfacial surface area then becomes a transition zone, although the thickness of the zone is variable. Steady flow then minimizes thickness and non-steady influences. Pumping, recharge, and tides increase the transition zone thickness. Transition zone thicknesses, once formed, range from

a few feet in undeveloped aquifers to many hundred feet in overpumped basins (Visher and Mink, 1964). The movement in the transition zone can also transport salt materials to the ocean. This circulation of flow has been verified by field investigations conducted in Miami (Cooper et al., 1964).

Control Methods for Saltwater Intrusion

Many methods have been utilized to control seawater intrusion (Todd, 1959). They include:

- *control of pumping patterns*—If pumping from a coastal groundwater basin is reduced or relocated, groundwater levels can rise and cause an increased seaward hydraulic gradient, so that a partial recovery from seawater intrusion occurs.

- *artificial recharge*—Seawater intrusion can be controlled by artificially recharging an intruded aquifer by surface spreading of freshwater or recharging wells causing water levels and gradients to be properly maintained.

- *maintenance of freshwater ridges*—Maintenance of freshwater ridges in an aquifer paralleling coastal areas creates a hydraulic barrier which prevents the intrusion of seawater.

- *installing extraction barriers*—By causing a reversal of the ridge method, a line of wells can be installed adjacent to and paralleling the coast and pumped to form a trough in the groundwater level causing a reduction in the usable storage capacity of the basin (California Department of Water Resources, 1970).

- *combining injection-extraction barriers*—Using the last two described methods, a combination injection ridge and pumping trough can be formed by two lines of wells along coastal areas with the total number of wells substantially increased.

- *forming subsurface barriers*—By constructing an impermeable subsurface barrier through an aquifer parallel to the coast, seawater would be prevented from entering the groundwater basin, although leakage due to the corrosive action of seawater or to earthquakes would need to be considered in barrier designs.

- *installing tide control gates*—Wherever drainage channels carry surplus waters from low-lying inland areas to the ocean, there is a danger of seawater penetrating inland areas during periods of high tides, and to control such intrusion, tide gates should be installed at the outlet of each channel to permit drainage water to be discharged to the ocean but prevent seawater from advancing inland (Parker et al., 1955).

Other Control Methods

A variety of other methods are available to control saltwater intrusion into freshwater aquifers (U.S. EPA, 1973). The selection of a particular method depends on local or regional circumstances initially responsible for the intrusion.

Control methods are described as follows:

- *Reduce pumping where upconing of saline waters occurs.*—Reducing pumping may take the form of actual termination of pumping or reduction in the pumping rate from individual wells, and the greater then is the tendency for the saltwater interface to subside and form a horizontal surface.

- *Increase freshwater groundwater levels.*—In situations where surface construction or excavations have lowered groundwater levels and caused underlying saline groundwater to rise, any action which raises groundwater levels is effective in suppressing intrusion.

- *Adjust pumping pressures because saltwater moves into a freshwater aquifer as a result of a pressure gradient.*—An effective control method is to reduce the pressure in the saltwater zone by drilling and pumping a well that has its casing perforated only in the saltwater portion of the aquifer (Gregg, 1971).

- *Have properly sealing wells to minimize vertical movement of saltwater in abandoned wells and test holes.*—Abandoned wells should be completely sealed with an impermeable material.

- *Construct wells in such a manner that they control the movement of saltwater within active fields that are either pumping or resting.*—During the drilling of a well, one or more zones of saline water may be encountered and those formations developed for freshwater production are selected with perforations placed within freshwater zones, and unperforated casing placed in saltwater areas (Campbell and Lehr, 1973; Gibson and Singer, 1971; Todd, 1959).

When freshwater aquifers need to be protected against vertical intrusion of saltwater, a water sampling network to monitor the effectiveness of the control method should also be incorporated. This monitoring system should consist of observation wells located so that perforations in the casing occur within freshwater zones. Water samples from such a system should be sampled regularly and analyzed for total dissolved solids and conductivity. The monitoring wells should be in the deeper portions of a freshwater zone to reveal evidence of any initial saltwater intrusion. These wells should also be spaced close enough to pumping wells, so that upconing will be detected. Again, abandoned wells or test holes should be tested to assure that they have been properly sealed. Regular measurements of pumping rates and groundwater level fluctuations would also help to recognize causal factors responsible for actual or incipient intrusion problems, if and when they are detected.

Monitoring for Intrusion of Seawater

A monitoring program is necessary as part of any system regardless of the method of seawater intrusion control applied. Conditions within and outside of intruded areas should be measured. Data collection within a systematic framework of groundwater levels is a must. The vertical structure of the transition zone should also be measured at more than one location.

Monitoring wells should be located in areas that provide an overall composite of the intrusion problems. These observation wells should be placed on the seaward side and the landward side. The number of wells required for any area will vary with problems pertinent to that vicinity. However, a reasonably dense network is usually required.

These wells should be measured for groundwater levels, and chloride and total dissolved solid concentrations should be determined from water samples. Sampling frequency should be dependent on fluctuations that occur. However, one to two months under normal circumstances should be considered a minimum. The observation wells constructed for the Los Angeles injection barrier (McIlwain et al., 1970; from U.S. EPA, 1973) were cased with four-inch PVC plastic pipe in a gravel-packed and grouted fourteen-inch diameter hole.

Saline Water and Its Presence in Inland Aquifers

Hydrologic data summarized indicate that large quantities of saline water exist under diverse geologic and hydrologic environments in the United States (U.S. EPA, 1973). In fact, much of the nation's largest inland sources of fresh groundwater are in close proximity to natural bodies of saline groundwater. Saline water derived from remnants of ancient sea inundations or water contaminated by natural mineral deposits occurs at shallow depths throughout the United States. Although, freshwater recharge has flushed much of the saline water, large volumes still remain at depths where the groundwater movement is restricted. Briny areas therefore occur in many areas and at various depths, especially in areas explored and developed by the oil industry.

Saline water in inland aquifers seems to be generally derived from the following sources (Task Committee on Salt Water Intrusion, 1969):

- that which entered aquifers during deposition or during a high stand of the sea in past geologic time
- from salt deposits in salt domes, thin beds, or dissemination of the material in geologic formations

- saline water concentrated by evaporation in pläyas or other enclosed areas
- return flows from irrigated lands
- saline wastes from anthropogenic sources

Salt Treatment of Roads and Streets

The contamination of groundwater resulting from application of salts for de-icing streets and highways in winter is a problem, which obviously occurs mainly in the northeast and northcentral U.S. (Hanes et al., 1970). Salt reaches groundwater both from storage stockpiles and from solutions that have been spread on roadways.

The New Hampshire Highway department also experienced long-term degradation of groundwater quality with highway de-icing salts (Hanes et al., 1970). Chloride content of water in certain shallow wells rose through time to con-

centrations of 3800 mg/l. This caused the casings and screens of wells to badly corrode to the point where they had to be replaced. Deutsch (1963) reported that a similar situation occurred in Michigan. There water from some wells contained 4400 mg/l of chloride as a result of highway salts.

Huling and Hollocher (1972) conducted an analysis of steady state concentrations of road salt added to groundwater in east central Massachusetts. Their study was based on an application rate of 20 metric tons of salt added per lane mile per year. It also took into account local rainfall and infiltration values. A chloride concentration of 100 mg/l was calculated for the area. Local deviations from this average value ranged from two to four times the figure of 100 mg/l, especially near major highways. The problem of road salt and its effect on increasing chloride levels is widespread and litigation on the matter seems to be common (Little, 1973).

ELEVEN

Contamination of Groundwater in Urban and Suburban Areas, and Diversion of Flow Structures

Effect of Urban Areas

Urban areas include suburban and central city complexes (Thomas and Schneider, 1970). More than two-thirds of the nation's population now reside in urban centers that occupy about 7 percent of the land area of the United States. By the year 2000 it is estimated that these areas will include three-fourths of the nation's population.

This concentration of people and their activities will produce tremendous demands on water supply systems and for treatment of residential, commercial, and industrial wastes. Water in some instances, for supply purposes, will have to be diverted and conveyed to urban areas, perhaps from sources hundreds of miles away. An example of this includes the Los Angeles–San Diego metropolitan complex. This complex receives water from the Colorado River and portions of Northern California.

Runoff and infiltration in urban areas are also markedly different than in undeveloped areas. Urban areas can produce hydrologic and hydraulic problems that significantly relate to development of water supplies (Schneider and Spieker, 1969; Rantz, 1970; Leopold, 1968).

Seawater intrusion in coastal aquifers is associated with urban areas that overpump, reduce natural recharge, and eliminate recharge from septic tank—leachfield systems, where public sewers have been installed. Runoff from urban areas, however, is also heavily contaminated, especially initial flows (Sartor and Boyd, 1972).

Urban leachate, as a source of groundwater contamination, owes its composition to dissolved organic and inorganic chemical constituents derived from a multiplicity of sources, i.e., dirty air, precipitation, leaching of asphalt streets, inefficient methods of waste disposal, lead in automobile exhausts, and improper disposal practices at numerous domestic, commercial, and industrial locations (Hackett, 1969). Urban pesticide use can also be a problem (Grier, 1981–82). Urban leachate can also directly contaminate streams or rivers, because many centers are located in lowlands adjacent to large streams. Groundwater withdrawals in urban areas may also cause contaminated water to flow from streams to hydraulically interconnected aquifers. Expansion of the populace from densely populated urban and suburban developments into rural or agricultural areas compounds problems of groundwater contamination by adding subsurface effluent from septic tanks or cesspool systems into groundwater areas. Additionally, in many urban and suburban areas, liquid wastes are accidentally or intentionally discharged on land surfaces and eventually reach shallow groundwater areas (Owe et al., 1982).

Urban areas in their mature stage produce large amounts of domestic, commercial, and industrial sewage, street runoff, and garbage (refuse). Contamination of groundwater from urban areas is not always confined to the immediate area where the contamination occurred. Effects can extend for considerable distances not only in groundwater areas but also in surface waters.

Ellis et al. (1986) studied sediment and metal loadings to roadside gully-pots (weight/area of catchment/runoff depth) for two defined sites within a highway catchment and compared the data to other urban studies. The higher metal loadings associated with the more active sites were particularly noticeable for Cu, but at both sites the relative sizes of the metal loadings reflected the expected metal availabilities. Stepwise linear regression analysis of pollutant loadings against five selected hydrological parameters indicated that total runoff volume and storm duration together explained over 90 percent of the observed variance in Pb, Cd, Mn, and sediment loadings. The derived model was used to predict pollutant removal rates over time, and these were shown to be consistent with those reported in other urban catchment studies.

Environmental Consequences

In urban areas, water quality can be degraded in shallow and deep aquifers. Increases in levels of nitrogen, chlorides, sulfates, and hardness of the water, have resulted in limitations on pumping from shallow aquifers in California and Long Island. High nitrogen content of groundwater in Kings County, Long Island, New York, at one time had been attributed to long-term leakage of public sewers (Kimmel, 1972). Contamination of shallow public water supply wells by cesspool effluent in Suffolk County, Long Island, New York, resulted in shutdowns of wells (Perlmutter and Guerrera, 1970). Similar problems have also occurred in California with salinity and nitrate problems in the Fresno-Clovis area (Nightingale, 1970). The potential effects of pesticides on aquifer(s) were discussed earlier (chapter 6).

Urbanizing alters the hydrology of an area and this results in a decrease in the natural recharge to underlying groundwater. Unless this recharge is compensated for by artificial recharge, it will have an adverse effect on groundwater quality, especially, if natural recharge levels were initially high (Seaburn, 1969). In urban areas, peak storm runoff and total runoff also increases over shorter time periods, and this results in a decreased streambed percolation rate. Streambed recharge is also further decreased by concrete storm drains and the lining of natural channels for flood control purposes.

In the Santa Ana River in southern California, contamination of groundwater resulted from the importation by municipalities of Colorado River water. Colorado River water is high in salinity (750–850 mg/liter and total dissolved solids), and contamination of groundwater resulted from use of the water to artificially recharge areas and also because it percolated into groundwater areas after irrigation of lawns, parks, and agricultural grounds.

High groundwater temperatures attributed to recharge of warm water used for air conditioning purposes were investigated in Brooklyn, New York, by Brashears (Brashears, 1941). Pluhowski (1970) in a similar study attributed a 5–8°C rise in summer temperature of water in Long Island streams to a variety of urban factors. These urban factors included pond and lake development, cutting of vegetation, increased stormwater runoff into streams, all associated with decreased groundwater flow.

Wikre (1973; from U.S. EPA, 1973) reported several incidents involving groundwater contamination related to urbanization in Minnesota. Drainage of surface water through wells via use of sumps produced discolored and turbid water. Positive coliform bacteriologicals, contamination from leachate produced by landfills, and solvents disposed of in pits and basins also caused major problems.

Sources and Nature of Contaminants in Urban Areas

Groundwater in an urban environment contains many inorganic and organic contaminants. Principal urban contaminants include the following (U.S. EPA, 1973):

- from atmospheric conditions: particulate matter, heavy metals, salts
- from precipitation: particulate matter, salts, dissolved gases
- from seawater encroachment: high dissolved solids, particularly sodium and chloride
- from industrial lagoons: heavy metals, acids, solvents, other inorganic and organic substances
- from cesspool, septic tank, and sewage lagoon effluents: sewage contaminants including high dissolved solids, chloride, sulfate, nitrogen, phosphate, detergents, bacteria
- from leaky pipelines and storage tanks: gasoline, fuel oil, solvents, and other chemicals
- from spills of liquid chemicals: heavy metals, salt, and other inorganic and organic chemicals
- from urban runoff: salt, fertilizer chemicals, nitrogen, and petroleum products
- from landfills: soluble organics, iron, manganese, methane, carbon dioxide, industrial wastes, nitrogen, other dissolved constituents, bacteria
- from leaky sewers: sewage contaminants, industrial chemicals, miscellaneous highway contaminants
- from stockpiles of solid raw materials: heavy metals, salt, other inorganic and organic chemicals
- from de-icing chemicals for roads: salts

The three most significant sources of groundwater contamination can include: industrial point-source discharge, municipal waste treatment plant effluent, and nonpoint source discharges (Schmidt and Spencer, 1986). However, urban stormwater runoff has not always been seriously considered as a major contributor of pollutants, even though under certain conditions it can dictate the quality of the receiving water (Hunter et al., 1979; Hoffman et al., 1982). Therefore, urban stormwater runoff quality must be considered and eventually controlled (National Pollutant Discharge Elimination System; NPDES Regulations will require this in the future).

At present, nonpoint efforts have focused on public education, recycling waste oil, and street sweeping as some of the activities. However, the results have been limited in scope. Direct connection of individual, commercial, and industrial pollutant discharges to storm drains is a major contributor of pollution and control of these sources is yet to become widespread.

Movement of Contamination

The mechanisms of groundwater contamination in urban areas include infiltration of fluids at or near land surfaces and leaching of soluble materials through the surface. Sources of fluids include their initial disposal through wells, pits, basins, and seepage from leaky stormwater drains, sanitary sewers, water mains, gas mains, steam pipes, industrial pipelines, cesspools, septic tanks, and other subsurface discharge facilities. Although natural treatment of some fluids occurs as they seep downward through various soil zones, large quantities of contaminants reach the water table. From there, contaminated water moves laterally toward natural discharge areas or pumping wells.

Control Methods for Preventing Contamination

The following list describes procedures that can be used to prevent, reduce, or eliminate contamination in urban and suburban areas (the applicability of any particular method being dependent on local circumstances and/or existing federal or state regulations):

- pre-treatment of industrial and sewage wastes before disposal into lagoons, or other approved receivable areas
- lining of disposal basins to prevent leaching into groundwater
- collection, by means of drains and wells, and treatment of leachate from landfills, industrial basins, and sewage lagoons
- proper management of groundwater pumping to prevent, stabilize, or retard saltwater intrusion into coastal freshwater aquifers
- creation, by means of installing wells, injection ridges or pumping troughs to retard saltwater intrusion
- abandonment or prohibition of on-site septic tank and leachfield systems in densely populated areas and replacement of sanitary wastes by addition of central or municipal sewer systems
- proper construction of new wells and properly abandoning wells no longer used
- implementation of adequate land storage of wastes and monitoring of industrial wastes and contaminants through permits, inspection, and enforcement activities
- reduction in use of de-icing salts for roads
- storing chemicals under adequate cover and on impermeable platforms to prevent leaching, and recovery of leachate if spillage occurs
- providing information for optimal applications of lawn fertilizers and garden chemicals to minimize potential leaching into groundwater areas
- frequent and adequate cleaning of streets with proper disposal of wastes collected
- artificially recharging selected areas with high quality water to compensate natural recharge use
- use of high quality water for municipal and industrial purposes where return flow from these uses enhances groundwater quality
- desalination of wastewaters before discharge
- providing for treatment of runoff from urban areas prior to discharge into streams essential for critical recharge areas

Where urban areas obtain groundwater from local wells for their water supply, the wells should be part of a monitoring and surveillance program that is associated with urban activities. Specific threats to groundwater quality should not have to occur prior to installation of monitoring wells to provide advance warning of potential contaminants approaching these essential water supply wells. Monitoring of its ambient quality for inorganics and organics, and more than those listed in the Safe Drinking Water Act (SDWA), is highly desirable to reduce or prevent further or long-term contamination.

Structures to Control Water

The construction and operation of structures (i.e., dams, levees, channels, floodways, causeways, and flow diversion facilities) can also influence groundwater quality (U.S. EPA, 1973). An awareness of them is necessary in order to control any possible resulting groundwater contamination.

Dams

An important effect that a dam can have on groundwater quality takes place where the foundation of the structure provides a complete cutoff of groundwater flow within an aquifer. Prado Dam on the Santa Ana River in southern California, a U.S. Corps of Engineers flood control structure, is located at the upper end of a narrow canyon which forms a natural outlet for surface and groundwater movement out of the Upper Santa Ana Valley (U.S. EPA, 1973). The cutoff wall in this instance extends to bedrock and blocks subsurface flow. This stoppage of subsurface flow can increase accumulation of contaminants in groundwater, and the natural disposal of salinity from the basin is also eliminated. The resulting accumulation of salts from natural or human-made sources (e.g., irrigation return flows) will through time also increase groundwater salinity.

Another effect due to placement of the dam involves the higher water table that is created behind the dam. This raises groundwater levels where the opportunity for contaminants from agricultural and septic tanks plus leachfield wastes can contact the water. Marshy areas, swamps, and pools are also created and evapotranspiration losses in turn concentrate salinity levels.

The reservoir created by the dam will also affect groundwater levels and quality in the area. If water is stored in the reservoir for long periods of time, the effects can be more pronounced than those resulting from construction of the dam. If the quality of the water in the reservoir is better than that of the groundwater, improvement in groundwater quality results. However, the opposite can also occur: groundwater quality can be degraded. Seepage losses from a reservoir storing poor quality water will also degrade groundwater quality.

Prevention of Contamination by Dams

Methods to control groundwater contamination by construction of dams should include a detailed evaluation of the following:

- designing of the dam and foundation to minimally restrict the down-valley flow of groundwater
- making provisions for controlled releases of water past the dam
- lowering of the water table upstream from the dam by placement of pumping wells
- minimizing or eliminating potential sources of contamination in the area upstream from the dam, possibly involving changes in land use, reduction in application of agricultural fertilizers or removal of domestic animals from the area
- selecting a site where seepage losses are minimal if the reservoir stores water of poor quality

Levees

Levees are low structures located along edges of surface water bodies (e.g., rivers, reservoirs, lakes) to prevent inundation of land behind them during periods of high water usually due to floods, storms, or tides. Levees are constructed to form controlled channels. However, only in rare instances do they form a barrier to groundwater movement.

Levees in coastal areas prevent the flooding of land by seawater. The quality of groundwater in the aquifers behind these levees is then protected. Harmful effects of levees on groundwater quality, however, do occur in floodplain areas of rivers (U.S. EPA, 1973). This is because the mineral quality of floodwaters is higher than that of groundwater for most areas. Under natural conditions during periodic inundations that occur in floodplain areas, water infiltrates to the groundwater and degrades its quality. If levees have been built in floodplains, the natural recharge to these areas is reduced. Consequently, the mineral quality of the groundwater will tend to concentrate through time.

To counteract the effect that levees have in degrading groundwater, two remedies should be considered. They include:

- pumping groundwater from the aquifer behind the levee to increase the circulation of groundwater and remove accumulations of salinity
- diverting freshwater to the land behind the levee

By over-irrigation or artificially recharging areas behind levees with water of a quality equal to or better than that present, a dilution similar to that produced by natural floodwaters can be achieved.

Channels

Channels are constructed to alter the alignment or configuration of water in natural rivers or streams. This is usually done to aid in navigation or to provide flood protection. Artificial channels also alter the natural circulation of the groundwater, and natural recharge to the groundwater may be increased or decreased depending upon the location, depth, and size of the channel. A thorough investigation of potential effects of channelization on quantity and quality of groundwater should be made before undertaking these projects.

Lined or unlined channels will have a different effect on groundwater quality (U.S. EPA, 1973). A lined channel, constructed of impermeable material (e.g., concrete, etc.) prevents the natural recharge of streamflow to groundwater.

To prevent the lowering of the groundwater as in the case of levees, water needs to be artificially recharged. Ditches or basins for artificial recharge can be installed in the vicinity of the lined channel. High quality water diverted from streams or derived from other acceptable sources and released close to these structures would infiltrate to groundwater levels and compensate the loss of natural streambed recharge that occurs due to channeling. This method is practiced in California.

In unlined channels, the water table elevation changes close to the channel. It becomes higher in areas close to the channel. If the channel is underlain by saline water, the reduction in freshwater head may cause the saline water to rise in adjacent areas and contaminate fresh aquifers. Methods to control this saline water rise include (U.S. EPA, 1973):

- installation of pumping wells in the underlying saline water zones and removal of this water by pumping and providing means for the disposal of saline water such as can be accomplished by evaporation from lined basins; disposal to the ocean, or desalting
- lining the channel with an impermeable material that will prevent dewatering of the upper portion of the aquifer and maintain original natural conditions of groundwater quality

The effects of unlined channels on groundwater quality are basically the same as for lined channels. Artificial recharge can be used to compensate for the loss, and unlined channels may also allow contaminated water to enter the groundwater if the level is below the bottom of the channel.

In some coastal areas (e.g., Florida and California) natural channels have been deepened and new channels excavated. However, these cut deeply into and through underlying clay formations which originally provided a natural barrier and prevented downward movement of saline water into freshwater aquifers. Groundwater contamination therefore resulted. Channels close to coastal areas should be located, designed, and constructed so natural barriers to saline water intrusion are not impaired. The channels should be lined with impervious material where this is not possible.

Bypass and Floodway Channels

Bypass and floodway channels are wide channels to carry floodwaters. These consequently are unlined and the bottom elevation is close to ground surface levels. The effect of such channels on groundwater quality often is minimal, particularly as they typically carry water for small fractions of each year (U.S. EPA, 1973). Floodwater flowing in a bypass channel that infiltrates tends to improve groundwater quality. However, because of the unknown effects of many waste products that can enter such systems, monitoring and/or observation wells should be installed at key locations. Specific control measures, if they are needed, can be directed towards problems identified in any monitoring program.

Causeways

A causeway is a raised highway or railway across low or wet ground. It is usually constructed of earthen embankments interspersed with bridged openings at systematic intervals (U.S. EPA, 1973). The principal effects that causeways can have on groundwater quality include disruptions of flow of surface water, which in turn could impair groundwater recharge or increase the percolation of contaminated water through ponding. The potential effects should be periodically monitored, although they are considered minimal. Proper drainage should be provided to eliminate ponding in as many instances as possible.

Flow Diversion Facilities

Flow diversion facilities consist of gates, locks, and weirs to regulate the distribution and levels of surface water. These structures are smaller than dams. Effects of these structures on groundwater quality are analogous to those previously described for dams. Similar control measures to protect groundwater quality can be adopted. Modifications would have to be made to accommodate the control procedures outlined for dams.

The operation of such structures and their effects on groundwater quality usually result from the diversion of water away from its intended course. Local recharge to the groundwater thus decreases. Ways to control this could involve controlled releases past the diversion structure. These releases should be sufficient in volume to provide the same amount of recharge that would have occurred. If this cannot be done, then artificial recharge may be practiced to compensate for the loss. Different groundwater laws in the various states might also have a direct relationship to quantities available for artificial recharge purposes.

TWELVE

Groundwater Contamination through Land Surfaces

Spills of Liquid Contaminants

Groundwater quality can be degraded as a result of discharges of liquid wastes from land surfaces. Discharges or spillages such as motor oil dripping from automobiles and trucks, toxic fluids released by careless handling, and various tank truck and/or car accidents, can cause this degradation. Tank trucks may habitually dump liquid wastes in open fields because adequate local regulatory authorities and enforcement actions to halt such incidents are usually lacking in most areas.

General considerations of hydrogeology make it logical to assume that in many instances groundwater quality beneath some industrial and urbanized areas is being gradually degraded by spills of various types of liquids on the land surface. Extensive chemical analyses of well waters that reveal such problems are now showing the need for establishment of comprehensive groundwater quality monitoring networks in these areas. Unfortunately, many areas do not have monitoring wells. Chemical analyses of water, when it is done, should cover not only a few of the common troublesome constituents, but should include those indicative of particular sources of contamination (e.g., organics, pesticides, heavy metals, etc.).

Contamination stemming from liquid spills can be quite variable. Degradation of groundwater quality, due to fuel spills and use of various types of cleaning agents have been detected at various places, including the Miami International Airport where petroleum-derived contaminants were floating on the water table (U.S. EPA, 1973). Widespread contamination of groundwater resources in Maryland were also traced to the disposal on the land surface of crankcase oils drained from cars and to oil spills from trucks, railroad tank cars, and oil drums.

The overall problem of spills of liquid contaminants can be severe in areas where high humidity occurs. Usually this includes the eastern, rather than the western, arid regions of the country, where low precipitation, low rates of groundwater recharge, and high rates of evaporation result in less opportunity for the percolation of liquid contaminants. Nevertheless, contamination of groundwaters from surface spilling have also occurred in the west, especially during large thunderstorms.

Environmental Consequences of Toxic Chemicals

Toxic chemicals spilled from trucks in road accidents or from tank railroad cars during a derailment can enter aquifers and move toward points of discharge (i.e., pumping wells or nearby surface water bodies). Contaminated water or liquids entering fractured rock (e.g., shales and granites) or rocks having large solution openings (e.g., limestones and dolomites) move faster than in different hydrogeologic environments. Adverse effects of contaminants entering fractured rocks or rocks having large openings that also provide water to municipalities some distance away from the area where the initial contamination entered, can take place quickly, and before the existence of the threat is recognized, problems can arise.

Municipal and private wells have also been contaminated by pickling brines and milk wastes discharged on the land surface to the water table in large industrial areas near Baltimore (Bennett and Meyer, 1952). Not only was the shallow aquifer contaminated by these wastes, but corrosion of the upper portions of the casings in deep wells allowed acid to move down into deeper aquifers. Contamination then showed up in nearby municipal wells.

Once an aquifer has been contaminated by substances derived from spills, groundwater quality remains affected for many decades, even if the source of the contamination is halted. Expensive control measures must then be implemented if the groundwater reservoir is to be purged of the contaminant, in addition to restricting the use of the aquifer and any surface body into which the contaminated groundwater discharges.

Movement of Contamination

In one-time spill situations, a large slug of liquid spread on ground surface areas can seep into the groundwater and remain intact. If long-term infiltration occurs, the contaminated fluid will move in a continuous fashion downward towards the water table and then become a plume stretching out from the initial source through the groundwater gradient. This action, once it reaches the aquifer, will degrade groundwater quality.

Composition, grain size, and thickness of soil materials in the unsaturated zone govern the movement of contaminated fluids moving toward the water table. The soil particles in the unsaturated zone filter some contaminants prior to them reaching the water table. During this process, when biodegradation will stabilize some organic fluids and an overall deficiency of moisture or low precipitation rate occurs, the contaminated fluid evaporates from soil particles or adheres to rock particles as thin films without descending to the water table. However, if the unsaturated zone is thin, little retardation, adsorptive action, or biodegradation occurs. The contaminant therefore arrives unchanged at the saturated zone.

Once contaminants enter the saturated zone area, its direction and rate of travel are determined by groundwater flow patterns. The contaminants then penetrate into the groundwater body or float on top of the water table. Through time the contaminated groundwater then moves toward natural discharge points or nearby pumping wells. If movement of groundwater is slow, the contamination remains in the vicinity of its initial source of release for many years or decades.

Control Methods for Spills

Deliberate spills of liquid contaminants should be controlled by regulation, inspection, surveillance, and enforcement action regarding the unwarranted release of waste fluids on land surfaces. Such wastes should be treated by the concern initially producing the product, or these wastes can be collected and processed at centralized or regional treatment centers, designed specifically for this purpose. Hazardous waste centers, transfer, and storage areas, for handling industrial wastes are becoming more available in urban areas.

Accidental spills cannot always be prevented; however, they can be minimized by adhering to the following (U.S. EPA, 1973):

- requiring that industries handling and transporting hazardous liquids maintain emergency facilities and trained personnel ready to respond to accidents

- requiring that notice be given to appropriate agencies before hazardous substances are transported between certain sites where risks are involved (Todd, 1973)
- that municipalities inform police, firemen, emergency crews, and state and local health authorities of the dangers that exist to groundwater areas if accidental spills occur, i.e., the information should also provide instruction on how to reduce the degree of contamination

Methods for intercepting and removing oil and their products from groundwater have been described by the American Petroleum Institute (1972).

Monitoring Procedures for Spills

Contamination sources from liquid spills can be located if investigations are conducted in areas where spills of noxious liquids have taken place. Monitoring wells can also be installed in these areas to locate the extent of contaminated groundwater.

Other monitoring approaches include expanding existing wells in the area that are routinely sampled by federal, state, or local agencies and to evaluate the significance of the sampling methodology and analytical techniques. When data obtained from these sources is verified, then the information can be included as part of any monitoring network. Long-term trends of such water quality constituents would be noted and additional investigations conducted if conditions warrant.

Groundwater sampling in a general sense is a complex undertaking (U.S. EPA, 1976b, 1985). Cost effective monitoring and sampling relies on careful planning and critical reading of the scientific literature. These activities will assure that the application of well placement, construction, sampling and analytical procedures results in the collection of high quality data. The information needs of each program must be recognized and all subsequent monitoring network design and operations decisions must be made in light of the available data (U.S. EPA, 1985). High quality hydrologic and chemical data collected in the detection phase of monitoring are essential to planning future activities. Our understanding of natural, accidental, or knowingly contaminated subsurface areas is incomplete.

Land Surface Areas and Groundwater Contamination

Elevational gradient changes in land surface areas produced by overpumping of groundwater can have detrimental effects on groundwater quality (U.S. EPA, 1973). A gradual collapse of the land surface can occur. The abrupt failure of the land surface due to the movement of material into underlying cavities can also occur. These physical changes or collapsed areas, also known as sinkholes, can serve as entry points for groundwater and affect water quality. Limestone terrains, especially, are prone to sinkhole collapses and once contaminated wastes are introduced into such formations they travel for large distances.

Sinkholes and their formation most often result from cavities in residual clay which have collapsed. The formation of the cavities may have initially been caused by a lowering of the water table, which resulted in a loss of support to clay overlying openings in bedrock material. Fluctuations of the water table against the residual clay base, downward movement of surface water through openings in clay, or increases in water velocity within a cone of depression may also cause the formation of sinkholes.

Sinkholes have been reported in Alabama, Florida, Missouri, Pennsylvania, and South Africa (Aley et al., 1972). Dewatering of areas for mining purposes was held responsible for causing several collapses (Foose, 1967). In Missouri, construction of sewage lagoons and surface water reservoirs have triggered collapses. Collapses, however, are not always related to human activities and some are associated with earthquakes and their aftereffects.

Increases in organic matter, bacteria, and colloidal material in suspension are released into an aquifer after a collapse occurs. Artificial introduction of poor quality water also occurs into the underground through depressed sinkholes. Groundwater tracer tests using *Lycopodium* spores were made in limestone aquifers in Missouri (Aley et al., 1972). Subsurface water movements via this method revealed as much as forty miles of movement in thirteen days. The potential contamination hazard of movement of contaminants through long distances in a relatively short period of time was thus demonstrated.

To minimize the danger of collapses attributed to sinkholes, pumping of water from limestone terrains should be regulated to stabilize groundwater levels and avoid extensive areas of dewatering, and extensive hydrogeologic investigations should be conducted at sites proposed for use as surface water reservoirs in limestone terrains.

Effects of Upstream Activities

Human activities upstream from recharge areas of a groundwater basin influence the quality of the underground water. Manmade influences include: changing lateral inflow, altering downward percolation, introducing contaminants directly into inflow or recharge water, and reducing the capacity of recharge areas, although some results of upstream activities benefit the quality of a downstream aquifer, e.g., regulation by upstream reservoirs and controlled releases downstream. Therefore, programs for controlling and abating causes of deterioration to groundwater quality must be formulated from an identification of the sources of contamination.

Causal Factors and Environmental Consequences

Control of groundwater contamination caused by upstream activities should include two significant geographic activities: those occurring at the surface overlying non-water-bearing formations where the effects of downstream groundwater basins and usage completely are related to the surface recharge areas of those basins, and those occurring on the surface overlying groundwater basins where interdependencies with downstream aquifers are related to inflow or surface recharge. Understanding the distinction between the two previously described activities is essential if measures to control adverse effects on groundwater quality are to be instituted. This is also because some of the upstream activities to be described in the following paragraph have the potential for affecting quality in downstream groundwater basins common to both categories.

Increased use of upstream surface water, with resultant evapotranspiration losses from streams contributing recharge to downstream groundwater basins, reduces the flow available for recharge and increases salt concentrations. Any sustained reduction in recharge rate will eventually cause deterioration of groundwater quality as a result of higher concentrations of minerals and organic contents in the remaining groundwater. Similar adverse effects also occur when intrusion of poor quality water replaces good quality recharge water.

Urbanization in upstream areas will eventually cause significant changes in streamflow, quantitatively and qualitatively. Runoff from urban areas and agricultural lands carries contaminants during small storms and in the initial phases of large storms. These contaminants could enter a downstream groundwater basin directly from surface recharge, or indirectly through surface lateral flow and percolation. Runoff characteristics from developed urban upstream areas are substantially different than from undeveloped areas. Higher peak discharges, greater volumes, and shorter runoff times occur in urbanized areas. Downstream percolation is thusly reduced and artificial recharge then becomes difficult.

Direct discharges of wastes upstream from the recharge area of a groundwater aquifer can be an obvious source of contaminants entering the basin as recharge waters. These discharges during certain time periods can include many waste generating activities, such as municipal sewage, commercial, and industrial wastes, sawmill and mining discharges, and drainage from feedlots.

Reduction of Groundwater Contamination Activities by Watershed Management

Watershed management practices involving changes in ground cover overlying a water supply aquifer may cause a significant rise to occur in water tables of the area. Nitrate contamination of shallow groundwater basins in Runnels

County, Texas and other sections of the state could have resulted from extensive changes in watershed management practices. Water tables rose in these areas and consequent leaching of near-surface rocks was dominant (Sopper, 1968). These watershed management activities may also cause a decrease in recharge rates by creating conditions where losses from evapotranspiration increase because of the fluctuating water tables. Farm ponds and stock-watering ponds which have been constructed in watershed areas have also significantly impaired downstream runoff. However, some watershed management practices, such as selective cutting and snowpack management in mountainous areas are beneficial in increasing recharge to downstream groundwater basins.

Fire, caused by either humans or natural processes, can have major upstream effects that impact directly on groundwater quality in downstream aquifers. Fire changes land surface areas, that in turn affect runoff rates and increase production, transportation, and deposition of sediment. The increased silt load and their movement are then deposited in recharge areas of downstream groundwater aquifers. In some instances, this can cause a decrease in infiltration capacity, sometimes affecting artificial recharge rates, and making it more difficult. When recharge does occur, an increase in organic loadings as a result of fire can be the consequence. The gradual decomposition of these organic materials and their continuous percolation to aquifer areas result in deterioration of groundwater quality. The necessity to abandon several walls can then be the end result.

In the case of groundwater aquifers which are dependent on recharge from other sources, effects of activities overlying an upstream basin may have beneficial or adverse effects. If the quality of water in the upstream aquifer is superior to that in the downstream one, pumpage from tributary groundwaters has a detrimental effect on the quality of the water, because the high-quality water entering the lower basin as subsurface inflow has been reduced. Conversely, if groundwater quality in the upstream aquifer is inferior to that of the downstream one, then water quality will be enhanced. Effective groundwater contamination control in upstream areas will alleviate adverse quality effects in downstream aquifers.

Control Methods to Reduce Upstream Contamination

Specific control measures to reduce the potential for groundwater contamination from activities in upstream urbanized or watershed areas include (U.S. EPA, 1973):

- adequate fire control efforts in tributary watersheds and prompt mitigative measures after fires occur, e.g., re-seeding immediately after a fire
- forest management practices, particularly in tributary watersheds that increase highquality water runoff

- land use controls in upstream urban areas including recharge zones to minimize threats to downstream groundwater quality—new institutional arrangements would be required
- treating urban runoff prior to its release, wherever possible or feasible
- managing interrelated groups of groundwater aquifers as integrated, with quality protection and maintenance being a principal objective
- artificially recharging aquifers with high-quality water in areas where natural streambed percolation has been reduced due to upstream use
- controlling releases from upstream reservoirs in such a way that adequate downstream groundwater recharge is maintained
- storing water from tributaries to facilitate recharge
- mechanically scarifying streambeds to maintain an adequate infiltration capacity

GROUNDWATER BASIN MANAGEMENT

Managing a groundwater aquifer can be considered analogous to the operation of a surface water reservoir. By regulating water within a dam, the reservoir can serve beneficial purposes. The benefits derived depend on varying water levels to meet supply and demand criteria.

Groundwater aquifers are now recognized as important resources for water storage and distribution, and they have numerous advantages over surface reservoirs. These advantages include (Comm. on Groundwater, 1972):

- The initial costs for storage of water are essentially zero.
- Siltation in aquifers when compared to reservoirs is not a problem.
- Eutrophication does not occur in underground reservoirs and is not a problem.
- Water temperatures and mineral quality once stabilized are relatively uniform.
- Evaporation losses are negligible.
- Turbidity is generally insignificant and surface area is not required.
- The life of an aquifer is often indefinite.

Groundwater basin management strives to provide a continuing supply of groundwater of satisfactory quality for the least amount of expense. However, to achieve this comprehensive geologic and hydrologic goal, development of mathematical models to simulate the aquifers, economic analyses of alternative operational schemes, and regulation of the aquifer are required. Conjunctive use of groundwater systems must also be considered in seeking maximum uses at minimum costs (Todd, 1959).

Investigations leading to proper management of an aquifer also consist of the following phases and components (U.S. EPA, 1973):

- geologic evaluation phase (e.g., data collection, water level maps, determination of storage capacity and change, transmission characteristics, water quality analysis, recharge rates)
- hydrologic evaluation phase (e.g., data collection, water demand, water supply and consumptive use, hydrologic balance)
- operational phase (e.g., future water demand, cost of facilities, operational plans)

- development of mathematical models for predictive purposes

The management of an aquifer involves evaluating patterns and schedules of recharge and withdrawals of water. This includes specifying the number and location of wells, pumping rates, and limitations on total withdrawal. Water quality objectives are set, and sources of contamination identified, controlled, and minimized. In some instances, artificial recharge or importation of water is also involved. Measures to limit saltwater intrusion should also be included. Regardless, however, a continuing data collection program is essential because of the dynamic nature of a groundwater resource system.

THIRTEEN

Groundwater Contamination: Ecotoxicology and Risk Assessment

Ecotoxicology of Groundwater

Ecotoxicology, a relatively new science, is concerned with toxic effects of chemicals and their relationship to living organisms within the area they occupy, e.g., a defined community or ecosystem. Ecotoxicology relates to specific toxic properties which interplay with biotic species occupying defined communities. It includes analyses of transfer pathways for specific chemical agents, including their movement within biological systems (Rail, 1985b).

The term ecotoxicology was first used by Truhaut in 1969 as a natural extension from the science of toxicology (Truhaut, 1977; Moriarity, 1983; Butler, 1978; Ramade, 1979). The science of ecotoxicology by definition, was meant to provide information on toxic chemical relationships within defined ecosystems in which organisms live. An ecosystem was meant to include "communities of living organisms together with their habitat and the interactions which occur."

One of the main differences between classical toxicology and ecotoxicology is that ecotoxicology is a four-part subject (Butler, 1978). Any assessment of the ultimate effect of an environmental pollutant must take into account, in a quantitative way, each of the following distinct processes involved (Truhaut, 1975; Butler, 1978):

- release of a substance into the environment—The amounts, forms, and sites of such releases must be known if the subsequent behavior is to be understood.
- the substance being transported geographically and into different biota, and perhaps chemically transformed, giving rise to compounds which have quite different environmental behavior pattens and toxic properties—The nature of such processes is unknown for the majority of environmental contaminants and the dangers arising from their ultimate

fate is complex, although the effects of certain chemicals have been well documented in recent years.

- The substance must eventually also affect certain target organisms (e.g., humans, animals, etc.) and for this to be assessed, the type of exposure that is to be manifested must be examined.
- assessing the response of an individual and/or groups of organisms to the specified (or perhaps transformed) contaminant or pollutant over an appropriate time scale necessary to do harm

For a proper groundwater ecotoxicological assessment to be made, the previously listed combination of steps must be evaluated in a quantitative fashion. Consequently, because of the need for quantitative precision, another facet to the previous steps is also in order. This facet should include a recognition that there is also uncertainty and possible error in the current understanding of the other four. The "best" estimates of the final effect of certain substances being released into the groundwater environment willingly or unwillingly should be based on state of the art scientific information. And we know that in the process of generating such quantitative data, a natural by-product includes the uncertainties in the steps themselves and varying confidence in levels of the results.

These summarizations of quantitative data, however, do eventually lead to an ordering of priorities for ongoing action and/or further research. Problems given the highest priority should have to do with extreme toxics present in certain pathways or those which involve considerable hazard.

The procedures for assessing steps to evaluate hazards continue to be developed. The process of combining the conclusions into an overall judgment, including an analysis of the uncertainties involved, however, is still in its preliminary form. Assessing ecotoxicological danger, environmental inputs, and their effects on the total ecosystem of humans is a difficult thing. However, the evaluation of "risk factors" can be used as a tool if they can be defined as "the expected frequency of undesirable effects arising from a specified (unit) exposure to a pollutant" (Butler, 1978). The quantitative relations between the exposure to a pollutant and the risk of magnitude or undesirable effects under specified conditions defined by environmental and target variables, have to be the criteria.

Risk, when considered in terms of environmental risk assessment, recognizes that no known general methodology is available for conducting an evaluation that includes both humans and non-humans or ecological receptors (Cornaby et al., 1982). In fact, the terminology in the literature is not always clear, suggesting that our views and knowledge of environmental risk assessment are still evolving. Assessing the risks of aquifer restoration alternatives is not easy. Selecting an effective remedial technique involves the balancing of the need to contain contaminants within acceptable levels against the costs associated with cleanup measures.

According to Canter and Knox (1986) there are two generally accepted approaches to risk assessment. The first approach includes utilization of criteria or standards for a pollutant and working backwards (i.e., utilizing intrinsic properties of the pollutant and aquifer to develop a listing of possible alternatives). The second approach analyzes the effectiveness of various alternatives and compares their resultant concentrations with a given standard. No matter which approach is used, a "criterion, standard, or acceptable level" is involved, and for the most part, data of this type are non-existent.

In order, also, to complete an assessment of risk, one must deal with the question of exposure (Walker, 1985; Farrara et al., 1984). The "exposure concentration" of a particular substance is the concentration to which humans, fish, and other organisms at some level in the food chain, or even the environment as a whole, are exposed to at a particular time. And, according to Haque et al. (1980), three basic elements are required to estimate exposure concentration. These include:

(1) Information on the source of the toxicant, including such items as production rate and release rate to the environment

(2) Characteristics of the toxicant that describe its ability to travel and react in the natural environment

(3) Data that can be used to estimate the population at risk, including occupational characteristics, medical surveillance data, and socioeconomic use habits

According to Farrara et al. (1984) the first two elements previously listed can be incorporated into mathematical models for estimating the transport and fate of the toxicant and, therefore, the concentrations that will appear in different sectors of the environment, i.e., the output can be a time-history of toxicant concentrations at various locations in the environment. Fararra et al. (1984) also mention that the last step in exposure assessment (step 3) coincides with the first step in hazard assessment (i.e., comparisons of the toxicity of a chemical with the exposure concentration). Consequently, the degree of hazard then deals not only with toxicity but also the degree of exposure. Highly toxic substances with no exposure are not hazardous, whereas mildly toxic substances with high exposure could be very hazardous.

Exposure of any life form to a toxicant can lead to intake by absorption, ingestion, or inhalation. The retention dose is specific to the toxicant, the pathway of exposure, the concentration of the toxicant in the pathway, and the age or sex factor of an individual. Age or sex factors are important in that they could represent different food habits, excretion rates, and other physiological characteristics that determine retention or excretion of a toxicant.

Acute or chronic dangers from toxic chemicals in groundwater must also be considered. Acute toxicity problems present a more easily evaluated problem in terms of the risk associated with contact and exposure. Chronic toxicity problems involve more complex evaluations. Therefore, in order to assess the effect of low-level exposure to humans, one must rely on established theories describing relationships between exposure and response mechanisms. One must also extrapolate the data in essentially two ways. The first is that there is a threshold level below which no response is observed, and the second is that there always is some response regardless of how small the exposure level is (i.e., zero response exists only at zero exposure). Since zero exposure in today's technological society may be neither practical nor economically feasible, it then becomes difficult to define a "safe level."

The science dealing with the study of pollutants in ecosystems (ecotoxicology) is still young, and the state of its knowledge, though increasing, is limited. Until this changes, however, difficult decisions regarding regulation of environmental pollution still have to be made. How risk assessment evaluations can be conducted for aquifer restoration purposes follows.

Risk Assessment

Much of the work addressing groundwater contamination problems has been conducted in response to the Comprehensive Environmental Response, Compensation and Liability Act, PL 95-5110 (known as CERCLA or "Superfund"). Superfund sites require the development of a "remedial action master plan" (RAMP). The purpose of the RAMP is to identify the type, scope, sequence, and schedule of remedial projects which may be appropriate in meeting an identified need. The RAMP was designed as an approach for developing an optimal solution for meeting a given need. The RAMP analysis involves consideration of:

(1) Environmental impacts

(2) Costs

(3) Risks

The main problem faced by all risk assessment techniques is that a large portion of the needed information, such as risk pathways or acceptable concentrations, is for all practical purposes, unknown.

It is also at this point that one should recognize that groundwater contamination restoration is not a distinct isolated hydrogeologic problem. The solution to groundwater contamination restoration problems requires involvement from many entities and disciplines. Managerial personnel, technical personnel, and remedial-related personnel from the construction industry must be involved. Institutional personnel from different levels of government, including on occasion individuals from international concerns, also have to participate. Multidisciplinary approaches involving teams and individuals from each previously mentioned category are needed. Hydrogeologists, environmental health scientists, epidemiologists, environmental engineers, scientists, ecotoxicoligists, chemists, geologists, risk analysts, and other staff personnel must work together.

Perspectives on risk assessment include many aspects, and with groundwater, concerns range from its contamination to its effects on human health. Approaches dealing with risk assessment and groundwater also cover considerations of selected chemical properties of materials to calculation of numerical indices and presentation of information on the probabilities and likely consequences of catastrophic events.

The principles of health risk assessment and management of toxic substances are understood, but application of these principles is difficult, because the hazardous waste management and risk assessment processes are complex, and optimal decisions require multidisciplinary approaches. Inter- and intra-agency communication are important. In fact, risk assessment decisions are not always made with all the facts, because staff is not always trained in every facet that is necessary for each assessment. And it is almost impossible to have all the information at hand to properly evaluate:

(1) *Hazard identification*—the qualitative evaluation of the adverse health effects of a substance(s) in animals or in humans

(2) *Exposure assessment*—the evaluation of the types (routes and media), magnitude, time, and duration of actual or anticipated exposures and of doses, when known; and when appropriate, the number of persons who are likely to be exposed

(3) *Dose-response assessment*—the process of estimating the relation between the dose of a substance(s) and the incidence of an adverse health effect situation

(4) *Risk-characterization*—the process of estimating the probable incidence of an adverse health effect to

humans under various conditions of exposure, including a description of the uncertainties involved

However, risk assessments may still be performed in response to either short-term (acute) exposures from toxic substances, long-term (chronic) exposures in which no immediate threat to the public health is apparent, or to combinations of these. Risk assessments may also require quantitative mathematical models to estimate exposure of the probability of an event or may require immediate action and both human health effects and environmental effects must be considered. Multiple-pollutant toxic substances make this type of an evaluation even more difficult. This is because of an increase in the diversity of chemicals and the uncertainties surrounding their assessment and management. Incomplete exposure assessments, lack of suitable monitoring methods, incomplete dose-response assessments, and the lack of available toxicological data on many chemicals contribute to this.

The following technical considerations, skills, and work tasks/activities are necessary when evaluating risks associated with toxic chemicals. The interplay and relationships with groundwater are evident.

- *Work Knowledge and Skills:* toxicology; environmental toxicology, ecotoxicology; chemistry, analytical, organic, biochemistry; engineering, environmental, safety, quality assurance; epidemiology; statistics, biometry, biostatics; medicine; mathematical modeling, dispersions, predictive capacity; hydrogeology; industrial hygiene; environmental health; risk assessment; ecology, biology, microbiology; cartography, physical geography; computer use, data management; explosives; sociology; physics; communications; and others.

- *Work Tasks/Activities:* environmental surveys; epidemiological studies; analysis and collection of data; geological, hydrogeologic, meteorological investigations; sampling and analysis of groundwater, air, soil, etc.; integration and blending of data; detection and reporting of problems; cooperation with other agencies; dispersion modeling; communication; risk assessment; inventory of materials located in area; monitoring programs (air, water, other); survey of human population and other biota in area; assessment of potential risks from exposure; engineering surveys; and assessment of potential release of toxics.

Hazard Ranking Information (HRI) and Site Rating Methodology (SRM)

The U.S. Environmental Protection Agency developed the Hazard Ranking System (HRS) as a method for ranking toxic material facilities for remedial action according to risks to health and the environment (Canter and Knox, 1986). "HRS" is a scoring system designed to address problems resulting from movement of toxic substances through many sources, including groundwater. It reflects the potential for harm to humans or the environment as a result of migration of a toxic substance away from a central facility by different routes (e.g., surface water, air, groundwater). The listing of some factors that are considered include:

migration, containment routes, characteristics of waste, hazardous waste quantities, targets; fire and explosion, containment, waste characteristics, quantity, targets; direct contact, accessibility, containment.

Those that concern the groundwater route include:

depth to aquifer, net precipitation, permeability, physical state, toxicity, waste quantity, distance to nearest well (down gradient), population served by groundwater, evidence of ignitability or explosivity, reactivity, distance to human population (sensitive environments—habitat), land use, and accessibility.

The HRS does not result in quantitative estimates of the probability of harm from a facility, but does provide a "rank-order" in terms of a potential hazard. It also evaluates:

- the physical containment of toxic substances
- routes by which releases would occur
- characteristics of harmful substances
- potential targets

The SRM (Site Rating Methodology) is being used to prioritize Superfund sites in terms of remedial actions. The SRM includes (Canter and Knox, 1986):

- a system for rating the general hazard potential of a site (Rating Factor System, RFS)
- a system for modifying the general rating based on site-specific problems (Additional Points System, APS)
- a system for interpreting the ratings in meaningful terms (Scoring System, SS)

The RFS is used for an initial rating of a site based on a set of thirty-one factors (Kufs et al., 1980; Canter and Knox, 1986). These include:

population within 1000 ft, distance to nearest drinking water well, distance to nearest off-site building, land use/zoning, critical environments, evidence of contamination, level of contamination, type of contamination, distance to nearest surface water, depth to groundwater, net precipitation, soil permeability, bedrock permeability, depth to bedrock, toxicity, radioactivity, persistence, ignitability, reactivity, corrosiveness, solubility, volatility, physical state, site security, hazardous waste quantity, total waste quantity, waste incompatibility, use of liners, use of leachate collection systems, use of gas collection systems, and use and condition of containers.

For each of these previously listed factors a four-level

rating scale was also developed (i.e., 0,1,2,3; with 0 indicating no potential hazard, to 3, indicating a high hazard). The scales were so defined so that they can be evaluated on the basis of information obtained from published materials and public records, or other sources of reliable information.

As was described, the rating factors do not assess the same magnitude of potential environmental consequence. Therefore, a multiplier is assigned to each factor in accordance with its relative magnitude of impact (i.e., a population within 100 ft of a site is rated higher than one over 3 miles).

Once a site has been rated using the SRM, the scores must be interpreted and this helps determine the order of sites for:

- collection of additional background information
- surveys of sites
- complete investigations of sites
- implementation of remedial action
- preparation of enforcement cases

Summary

Risk assessment should focus on prevention activities that will offset long-term effects of groundwater (or surface water) contamination. Preventive activities such as underground storage tank and underground pipeline evaluations prior to installation, all contribute to the ultimate goal of protecting groundwater supplies. Preventive programs are also going to involve many disciplines, not only hydrogeologists. Individuals working in this area must possess information or have expertise in:

hazardous waste regulations and programs; hazardous waste assessment methods and techniques, survey techniques, sampling techniques and strategies, enforcement procedures/legal support, environmental toxicology (ecotoxicological concepts and principles), epidemiological investigations, site assessment and remediation, personal protection and safety, health interactions, and other related concerns.

Included also will be information that needs to be generated on groundwater protection from facilities that treat, store, or dispose of hazardous waste in surface impoundments, waste piles, land treatment units or landfills; 40 CFR Ch. I (7-11-85 Edition):

- the physical and chemical characteristics of the waste in the regulated unit, including its potential for migration
- the hydrogeological characteristics of the facility and surrounding land
- the quantity of groundwater and the direction of groundwater flow

- the proximity and withdrawal rates of groundwater users
- the current and future uses of groundwater in the area
- the existing quality of groundwater, including other sources of contamination and their cumulative impact on the groundwater quality

- the potential for health risks caused by human exposure to waste constituents
- the potential damage to wildlife, crops, vegetation, and physical structures caused by exposure to waste constituents
- the persistence and permanence of the potential adverse effects

FOURTEEN

Groundwater Contamination: Its Management and Prevention

Proper Management

Proper management of groundwater aquifers to eliminate or control groundwater contamination, requires appropriate institutional structures. The aquifer should be managed to insure that water quality is not detrimentally affected on the short- or long-term basis. Management for control purposes should include:

- maintaining groundwater levels so as to minimize the opportunity for contamination of infiltrate from surface sources
- maintaining groundwater levels to prevent upward movement of more saline and warmer water into the aquifer
- regulating the quality of water used to artificially recharge the aquifer—Storm runoff collected in upstream reservoirs, stored, and then released into spreading areas could be of a higher quality than groundwater; however, imported and reclaimed waters may not be.
- preventing saltwater intrusion and inflow of poor quality natural waters from adjacent surface areas and aquifers—Poor quality water from underground sources can usually be excluded by many pumping wells installed in a line, while surface waters can be intercepted by drainage ditches and diverted from the area.
- regulating the drilling, completion, and operation of all types of wells penetrating the aquifer in question
- reducing salt loads by exporting groundwaters, wastewaters, or brines high in salinity
- systematically monitoring the quality of groundwater throughout the aquifer to identify and locate contamination sources (leaking underground fuel tanks, etc.) or to verify if corrective measures have been successful

- proper implementation of comprehensive planning programs aimed at controlling and abating groundwater contamination (i.e., such as comprehensive pre-treatment programs for industrial waste discharge).

Protection of Public Water Supplies

The protection of groundwater used for public supply is dependent on an adequate understanding of the fundamentals of groundwater hydrology (U.S. EPA, 1985). However, the subsurface environment is a complex system subject to contamination from a host of sources, as have been described in this book. Furthermore, the slow movement of contaminants, when they do occur through the groundwater environment, results in longer residence time and little diffusion. The restoration of groundwater quality therefore becomes difficult and expensive (Anonymous, 1981). Restoration costs usually exceed short-term value concerns of the resource and the most viable approach to groundwater quality protection must be one of the prevention and not cure (Morrison, 1981).

Protecting groundwater resources from contamination requires many different "best management" practices, including the development of plans at the local level to control activities that threaten the resource (Josephson, 1985; Wise, 1977; Rail, 1985a).

Adequate prevention, planning, and emergency response (U.S. EPA, 1983) may be beyond the resources of water utilities and it therefore becomes the responsibility of the local community to carry out planning and preventive programs that will assess drinking water needs and protect present and future water supplies. Suitable aquifer protection controls must be adopted under local planning and zoning laws which take into account land use, industrial development, health, housing and agriculture. No one approach would successfully protect all aquifers. The entire community, including local government, is responsible for balancing the risks, the costs, and the benefits involved in protecting the groundwater supply.

Community planning is evident in some states (e.g., Arizona, California, Massachusetts and Cape Cod, Colorado, Connecticut, Florida and Dade County, Kansas, New York, Long Island, New York, New Jersey, Wisconsin; NRC, 1986b; Moorehouse, 1985). The National Research Council of the National Academy of Sciences (NRC, 1986b) identified several state and local groundwater protection programs, focusing on prevention of groundwater contamination with respect to their scientific bases, performance over time, administrative requirements, and their legal and economic frameworks. The resulting report by them summarizes the committee's review of case studies and identifies those significant technical and institutional features that show progress and promise in providing protection of groundwater quality.

Included in their write-up (NRC, 1986b) are chapters concerning:

- background information on groundwater protection strategies
- groundwater quality standards and contamination sources
- summaries of the state and local groundwater programs reviewed
- state and local strategies to protect groundwater

Groundwater Contamination Control

Methodologies for groundwater quality protection or treatment depend on whether the contamination problem is acute or chronic (Canter and Knox, 1986). Acute contamination may occur from inadvertent spills of chemicals or releases of undesirable materials and chemicals during a transportation accident. Acute contamination events are unplanned and are characterized by their emergency nature. Chronic aquifer contamination may occur from numerous point and area sources and may involve traditional contaminants such as nitrates and bacteria, or unique contaminants such as petroleum fuels (benzene), metals, and organic chemicals.

It should also be recognized that a given aquifer cleanup project may involve usage of several methodologies in combination, such as excavation, backfilling, transportation of wastes to hazardous waste disposal sites, contaminant removal wells, treatment of contaminated water, discharge to municipal or surface drainage, surface capping, subsurface barrier installation, and in situ chemical treatment. Cantor and Knox (1986) discuss many of these aspects in their groundwater pollution control text (e.g., physical control measures, treatment of groundwater, in situ technologies, aquifer restoration, decision making, risk assessment, public participation, case studies).

Establishment of Groundwater Protection Programs at the State and Federal Level

State and federal programs for protecting groundwater quality, including detecting, correcting, and preventing contamination, must continue to be established and are continually being expanded (Bacon and Oleckno, 1985). Consequently, many of these efforts have made significant contributions to the protection of groundwater. In some states, major sources of contamination have been identified, inventories have been conducted, incidents documented, and scientific advances made in understanding groundwater hydrogeology of their respective areas.

At the federal level, many statutes now authorize programs relevant to groundwater protection, and more than two dozen agencies and offices are involved with groundwater contamination related activities. Most states are now concerned about contamination and have programs, at varying stages of development, to protect their groundwater. However, despite expanding federal and state efforts, programs are still limited in their ability to protect many areas against groundwater contamination. There is no explicit national legislative mandate to protect groundwater quality, and although the groundwater protection strategy of the U.S. EPA acknowledges the need for comprehensive resource management at the federal level, details have not been provided. Significant efforts, however, are being made with the amendments to the Safe Drinking Water Act (SDWA), and include Wellhead Protection grants. Most authorized programs are also in early stages and some are still a long time from being fully in place. In addition, groundwater quality programs are not always coordinated among the proper agencies and in some instances they are not consistent with other programs for groundwater quantity or surface water.

From a groundwater protection analysis, some federal and state programs have not provided a proper focus in reference to sources, contaminants, and users. Most programs involve managing selected sources of groundwater contamination, analyzing for selected contaminants, and overseeing drinking water in public water supply systems. As a result, many states, in addition to federal legislation have their own standards for drinking water and groundwater quality.

PROGRAMS FOR CORRECTIVE ACTION OF GROUNDWATER CONTAMINATION PROBLEMS

Few comprehensive corrective actions to solve groundwater contamination problems in the United States have been undertaken to date, relative to the number of sites identified as requiring action. Although federally funded corrective actions authorized by the Comprehensive Environmental Response, Compensation, and Liability Act (CERCLA), address sources and contaminants, the actions are limited to hazardous waste disposal sites. Corrective action, in many instances, has also not involved cleanup of contaminated groundwater.

State corrective action programs, in many instances, are also at an early stage of development. Presently, a great number of programs relate to accidental spill situations and detection of leaks from underground storage tanks (USTs). Other state programs are designed either to retain (e.g., in landfills) or discharge (e.g., via municipal sewerage, injection wells, etc.) contaminants into surface or subsurface areas. Many state corrective actions still result from politically motivated complaints rather than systematic efforts to identify, evaluate, and control contaminated sites.

Few sources are addressed in federal and state programs to prevent groundwater contamination when one considers the full magnitude of the problem. State and federal programs focus on sources associated with toxic materials, although implementation and enforcement of many programs are still in infant stages.

Some approaches used to prevent groundwater contamination by some states have included provisions for evaluating design, operational aspects, siting, uses, and closing of point-source contamination areas. In most instances, remedies in this category have been mandatory. Additional approaches to prevent groundwater contamination include implementation of alternatives to a specific contaminating activity (i.e., make process or product changes that reduce waste volume, initiate waste recycling and recovery, pretreatment on the site).

Focusing on a specific source is an approach to prevention of groundwater contamination. Unfortunately, other types of approaches have not been extensively applied to groundwater. Few efforts have been initiated to control cause and effect activities located in recharge areas. It seems that if approaches are not source-specific, they carry no merit. However, the Federal Government does provide support for protection of selected recharge areas through the Sole Source Aquifer Program and Wellhead Protection provisions under the Safe Drinking Water Act and its recent amendments (Amendments to the Safe Drinking Water Act, 1986). Selected recharge areas are also being protected by certain states and local governments. This they accomplish through land use controls and land acquisition.

Another tack that should be implemented to prevent groundwater contamination involves placing restrictions on the manufacturing, generation, distribution, and use of specific contaminating substances. For example, pesticides may be introduced from a non-point source such as a specific land application, from a storage tank, from landfills, and from residential disposal. Although the Toxic Substances Control Act (TSCA) and the Federal Insecticide, Fungicide, and Rodenticide Act (FIFRA) authorize regulation of potential groundwater contaminants in this instance, application of programs addressed to correct or prevent groundwater contamination are limited. They must be expanded and enforced.

FOCUS OF PROGRAMS FOR PREVENTION OF GROUNDWATER CONTAMINATION

The effectiveness of federal, state, and local programs to protect groundwater from being contaminated has not only had a limited focus, but technical factors to solve problems have not always been applied. For example, hydrogeologic investigations are needed to detect existing problems, evaluate the performance of corrective actions, and monitor the effectiveness of preventive activities. The technologies for obtaining adequate hydrogeologic information are available. However, there will always be a certain amount of uncertainty about groundwater contamination investigations because of difficulties in dealing with indirect observations. Although many scientific advances have been made to improve the gathering of data and reliability of results, they also often increase costs and time required to conduct an adequate investigation.

There are additional major constraints on certain technical hydrogeologic investigations. For example, the technology for conducting adequate investigations in some geologic environments (e.g., fractured rock) occurring in many areas of the United States, is limited. But it should be understood that many hydrogeologic studies are costly, time-consuming, dependent on site conditions, and relate to the level of detail required in the investigation objectives. The reliability of hydrogeologic investigations also depends on the degree of skilled personnel that must tailor their study to the site-specific nature of a specific groundwater contamination problem. Unfortunately, it is in this area that properly educated personnel are in short supply.

In addition, many detection activities are constrained by the high costs of monitoring. Institutional limitations exist in some states, for instance, a lack of authority to obtain data about particular sources of contamination, such as private systems, or on federal Indian reservations. It must be recognized that water quality data is also difficult to analyze and interpret, especially if trace levels or mixtures of contaminants are present, or if changes in chemical and biological observations occur frequently.

Major constraints that limit corrective action on solving groundwater contamination problems also include: uncertainty about some techniques that can be used to improve groundwater quality, the high costs of implementation, the need to design measures appropriate for site-specific areas, and the lack of adequate base-line data. An exact definition of the nature of contaminants for a specific site is another constraint. Treatment cost and techniques are dependent on the contaminants present, and cleanup activities are uncertain when complex mixtures of contaminants and/or concentrations change through time. To date, correction alternatives, containment, withdrawal, treatment, in situ rehabilitation, and management options are selected, based on how rapidly they can be implemented, how effective they are, and the extent to which the uncertainties inherent in their performance can be reduced. Whether there is clear authority to implement any selected strategy is also important.

Corrective actions relate to site access, alternatives available for other methods of disposal, and being able to implement and carry through correction activities. Side-effects, however, can also have environmental consequences. For example, closing wells results in the continued presence of contaminants, and excavations to remove contaminants causes them to have to be transferred to another site where the problem will, hopefully, not be repeated.

Prevention efforts generally are retarded by lack of financial aid to implement programs, unresolved questions about the technical adequacy of certain available methods, or an incomplete understanding about the complexities of groundwater contamination. In some instances, political decisions can also be blamed for the problem.

NATIONAL POLICY IMPLICATIONS

National policy options relate to development and implementaton of federal, state, and local programs to protect groundwater. The federal framework has the potential to protect the nation's groundwater from additional contamination, but its realization is dependent on broadening the coverage of authorized programs. Implementation at all levels also requires that activities be coordinated among and within agencies. Political judgments concerning the role of the federal government and the importance of the states making positive progress in their abilities to detect, correct, and/or prevent groundwater contamination are also critical.

The development of national policies related to the protection of groundwater from contamination must also include recognition of site-specific problems for given areas. After contaminated sites are identified, adequate efforts to detect, correct, and prevent contamination must be designed to solve the problems at that site. National policy must then be flexible and be able to accommodate the various groundwater problems encountered by varying site conditions. The federal government must provide adequate support to the states for detecting, correcting, and preventing groundwater contamination. The government should be instrumental in providing funding, technical assistance, demonstration projects, and research and development.

Current federal laws and programs have assisted states in abating their groundwater contamination problems; however, the level of federal support to the states is still not adequate. An explicit national legislative goal to protect groundwater quality is necessary. Federal programs must address specific problems that vary from state to state.

GROUNDWATER TECHNICAL ASSISTANCE TO STATES AND LOCAL GOVERNMENTS

Technical assistance to the states must include training programs, guidelines, document distribution, and information exchange. This is because at the state or local level, qualified personnel are limited. If this is not corrected, then the nation's ability to protect its groundwater quality will suffer.

In addition, federal funding for training and education is necessary to achieve an increase in the technical capabilities of the nation to deal with the groundwater contamination problem. The U.S. Geological Survey Cooperative Program to state and municipalities and other related technical assistance programs must continue. Certification programs by the federal government, the states, municipalities, or professional societies must also occur and continue to ensure that personnel possess minimum technical qualifications.

Since the solving of groundwater contamination problems requires site-specific evaluations, they still have common features amenable to the development of federal guidelines. From a national perspective, the goal of these guidelines should then be to ensure that a minimum set of considerations are used for protection of groundwater quality. These guidelines or considerations would also be a way of providing information required by states or municipalities in handling their groundwater contamination problems. General guidelines could be developed for assisting states in evaluating priorities for allocating their resources.

In summary, the federal government could provide assistance to states in the following areas (OTA, 1984):

- *with respect to detection:*
  - guidelines to assist in conducting reliable hydrogeologic investigations, including monitoring of the flow system, sampling and analysis, and data interpretation
  - guidelines for addressing contaminants for which there are no federal standards
  - guidelines in setting priorities and determining which contaminant sources to monitor and inventory
- *with respect to correlation:*
  - guidelines in selecting and implementing corrective action when necessary
  - guidelines for cleanup standards determined on a site-specific basis
- *with respect to prevention:*
  - guidelines for preventing contamination from contaminating sources, presentation of alternatives for reducing the wastes generated by a source, and information on waste recycling (i.e., as part of preventing contamination from sources)
  - guidelines for considering protection of aquifer recharge areas, wellhead protection, proper land use, and for establishing a balance that enhances water quality

There are also several ways that the federal government can facilitate information exchange among the states and municipalities. It could provide information about the different state approaches to protection of water areas and would assist them in learning from successes and failures of each other.

RESEARCH AND DEVELOPMENT NEEDS

Research and development activities provide information that would support states in their efforts to protect groundwater. Key elements of research activities should include:

- *detection:*
  — research on toxicology and adverse health effects of contaminants that occur in groundwater, with emphasis on synergistic effects of mixtures
  — development of water quality standards for substances known to occur in groundwater that are not now covered; these standards could be applied to drinking water supply and groundwater quality programs
  — research on environmental and economic impacts of contamination
  — research on development of reliable techniques for conducting hydrogeologic investigations
- *correction:*
  — research on the behavior of contaminants in groundwater, specifically, chemical and biological transformation of organic chemicals
  — research on development of techniques for treating water contaminated from multiple sources
- *prevention:*
  — research on mechanisms for preventing contamination, including ways to reduce generation of items and disposal volumes

The protection of groundwater from contamination will depend on educating the public and law makers. All segments of society need to understand the problems and solutions. Then, they should all work together for a common cause.

Establishment of Groundwater Protection Programs at the Local Level

Good quality water is essential for health, aesthetic, and economic reasons. Adequate quantities of a supply are also important considerations. Comprehensive groundwater monitoring data bases (i.e., including quality and quantity parameters within an aquifer) are therefore important. Data bases are valuable, not only for preventing direct contamination of underground aquifers, but in applying corrective measures when problems have been identified. How to establish systematic groundwater monitoring network plans and programs for any underground water sources to avoid or identify water related problems can differ from one place to another. However, some ideas and recommendations on how to evaluate the many complexities that arise when working with a valuable resource such as groundwater are now presented. Suggested local, federal, and state government roles are also discussed. The emphasis of the discussion on developing a groundwater monitoring plan and program is placed at the local level, while at the same time it is recognized that implementation of such a work plan must often come from state or federal sources.

Protocol for Protection Programs—Key Roles of Implementation

Establishment of groundwater monitoring plans and programs within an aquifer are essential to determine if significant groundwater contamination has or has not occurred within a given area (Rail, 1985a). Completely "thought-out" plans and programs for an aquifer or a series of adjacent aquifers are necessary and should, as a minimum, include evaluations involving water quality (biological, chemical, radiological, etc.) and water quantity (water flow, water level, recharge rates, etc.).

Federal, state, and local government entities have also recognized this need for adequate plans and have always attempted in some way to manage and protect critical water resources to a limited degree. However, on the whole, they have been very slow in developing comprehensive programs of groundwater contamination monitoring, analyses, and prevention, although some states such as Oklahoma, Connecticut, and New Mexico have taken a lead in this direction. What appears to be missing for an implementation of programs in many states or municipalities is an ongoing groundwater schematic that provides flexible guidance to parties interested in setting up monitoring and protection programs. Monitoring and protection programs can then be used to protect and prolong the useful life of domestic groundwater areas or solve problems if they are detected.

Although, in general, basic groundwater monitoring related data efforts have been conducted in most states as part of the Safe Drinking Water Act, to establish a proper and essential monitoring and protection program, the following information needs to be addressed on a long-term basis. These initial plans and programs should also include information pertinent to water quality and quantity concerns of any given area and could encompass, as a minimum, the following basic steps (Rail, 1985a):

I. Select and Assemble Participants
II. Establish Purpose and Development Plans
III. Technically Evaluate the Aquifer(s), Including:

    (1) Identifying contamination potential

(2) Use of applicable available technologies
(3) Determine present and future demands on the aquifer(s)

IV. Select Computer Models for Data Summarization
V. Evaluate Water Law and the Responsibilities of Municipalities and Government
VI. Prepare and Implement the Plan and Programs

These steps are not listed in chronological order and can be addressed simultaneously depending on the geographical area being evaluated and its water concerns, etc.

Recommended Steps

I. SELECT AND ASSEMBLE PARTICIPANTS

Implementation of a groundwater monitoring and protection plan should depend on a high degree of involvement of those who will benefit from the resource. This is also the idea proposed in a local ordinance concerning groundwater in southeastern Minnesota (Gass, 1985; Hale et al., 1965). It is the beneficiaries (i.e., the local owners of the resource) whose future depends on continued availability of good quality groundwater for its intended use, whether it be used for agriculture, domestic, municipal, or industrial needs. Consequently, local interests should be used to guide and formulate water policy for the area.

All entities that utilize or have the potential to use water from the aquifer which is to be monitored or protected should be identified and encouraged to provide input into any development plans. In reality, however, it should also be recognized that sometimes area-wide concerns might still best solve the problems, particularly if local interest groups cannot agree on water priority issues.

Step I can possibly be accomplished by the formation of an Aquifer Water Steering Group made up of members from the local or water basin area in question (e.g., farmers, developers, old folks who knew what it used to be like, local, state, and federal agencies, etc.). The main function of the group or representatives from each area would be to provide input on developing groundwater monitoring and protection activities for the area. The steering group membership should also include water quality and quantity professionals. In New Mexico, for example, active members on the steering group could come from adjacent Indian reservations, the U.S. Geological Survey, the state Environmental Improvement Agency, the state Scientific Laboratory System, the state Engineer's Office, the state Soil and Water Conservation Division, Council of Governments, New Mexico Conservation Division, New Mexico Game and Fish Department, New Mexico Bureau of Mines, the Interstate Stream Commission, the U.S. Forest Service, the Corps of Engineers, the city of Albuquerque Public Works Department, Water Supply and Liquid Waste Divisions, and

county Environmental Health Department, state universities, military bases, and other related entities concerned with water. Membership in the group could also include a limited number of officials or unofficials from adjacent cities, towns, and communities that have interests in, and share the same aquifer(s).

Identifying participants and maintaining individuals and agencies that want to contribute their water expertise is not an easy task, although once the steering group has been established it becomes easier, since it determines the directions the group will take.

II. ESTABLISH PURPOSE AND A DEVELOPMENT PLAN

The responsibility for developing groundwater monitoring protection plans has to be given to someone. A professional staff should be retained to perform the investigations, if members of the steering group do not provide this function. The professional staff that is retained could in some instances consist of outside consultants which would have data available and accessible to them by the various entities that are participating in the steering group. However, whether a professional staff is contracted, hired directly, or consists of active members from the steering group (whose agencies and membership might be willing to allow in-kind services), eventually, they must also be supported by some means. Financial assistance for professional staff might have to be sought from state legislatures, city or county governments, and/or federal grant funding sources. A possibility of increasing various tax bases in the area and adjusting water rates within a specified district could also be explored as a source for generating revenue to support the activities of a groundwater monitoring and protection staff. The steering group could also investigate other possible funding sources and make recommendations for implementation.

Predictable and actual sources of revenue or in-kind time sharing allocations from the different members and agencies of the steering group that can be used to set up and maintain the eventual monitoring network must be identified early in its formulation. In-kind time sharing and the responsibility for developing certain phases of a monitoring network and protection strategy should not be made by members or agencies of the group, unless it is felt they can actively participate in the development and implementation process of a proposed groundwater monitoring and protection plan and program.

III. TECHNICALLY EVALUATE THE AQUIFER(S)

One of the first objectives of any proposed groundwater monitoring and protection effort should be to identify the boundaries of the aquifer(s) in question. A critical hydrologic analysis is essential if "internal" factors which cause significant changes, not necessarily detrimental, in the aquifer(s) are to be identified and described. Compilation of historical water quality data should be assembled from available sources (Rail, 1986; Unpublished Manuscript; See Appendix A). In New Mexico, for example, water related information can be summarized from a review of federal, state, and municipal records. Other states could also seek information from some of the federal agencies, such as U.S. Geological Survey representatives in their area. The agencies and municipalities that would provide water related information would differ in each state, but be similar to the ones in New Mexico. Other information concerning water related aspects of the aquifer(s) in question should also be summarized. These are discussed in more detail in Step IV.

After water data, as gathered in the preceding step for the aquifer(s), have been compiled and summarized, gaps or absence of data, must then be identified. If many gaps exist in the data from specific geographical areas or geological formations (including the quality and yield from various well depths), then an additional plan for procuring needed information is necessary. The possible selection and construction of new well monitoring sites, and the proposed location and identification of present wells which could be incorporated in any monitoring network must be included. However, monitoring wells should not automatically be drilled, unless they are needed or will provide important data.

After existing wells from the aquifer have been selected for monitoring purposes and new wells drilled, then water quality or quantity information should be gathered on systematic time-frame intervals (e.g., monthly, bimonthly, seasonal, yearly, etc.). The time intervals chosen should vary with the tests being conducted, i.e., qualitative, Safe Drinking Water Act parameters, Priority Pollutants (U.S. EPA, 1979), quantitative water levels, water movement, etc. The schedule should include intervals that are sensitive enough to detect significant and statistical changes within the water measurements taken (Nie, 1983; Steel and Torrie, 1960).

Water supply wells used for monitoring purposes, especially those belonging to municipalities, should (if costs are not too high) constructually modify their systems to provide accurate and reliable water samples and evaluations. Municipal water supply wells which are in the process of being constructed, or yet to be drilled, should be completed in such a way that they also meet the necessary specifications for their use in an ongoing groundwater monitoring program.

(1) Identifying Contamination Potential

When naturally occurring sources of water quality degradation (e.g., high nitrates or chlorides) within an aquifer(s) have been located and verified, past, present, and projected human activities should be described, especially, activities that could continue to be detrimental to water supply areas. State or local comprehensive plans should also be reviewed. Historical maps and aerial photographs should be examined, if available for the area. Potential pollution threats to groundwater quality, such as leakage from underground fuel tanks, location of abandoned wells from domestic systems, abandoned wells from oil and natural gas field exploration activities, large pits with high infiltration potential, injection wells, ponds, lagoons, landfills, feedlots, septic tanks and leachfields, disposal of septage, industrial wastes and other potential sources of contaminants to the aquifer, should be located and mapped. Stormwater runoff drainage and ponding areas resulting from urban and industrial sources should also be mapped and evaluated along with other potential recharge areas to the aquifer(s).

After a complete listing of potential and actual contamination sources to the aquifer(s) has been compiled, a Ranking System (RS) must be developed. This ranking system will be used to provide information concerning where the greatest potential for contamination occurs in the aquifer(s). This contamination potential RS, when compared with area land-use and other proposed planning and development plans, should help to minimize future contamination of critical areas (e.g., future municipal water supply well sites, wellhead areas, recharge zones, etc.).

Future development and expansion of municipalities into critical recharge zones can also be planned in such a way that the impact can be minimized through proper zoning and enforcement of requirements. Incorporation of open space, preservation of natural habitat, and/or use of water detention facilities (i.e., if acceptable to Water Law and Water Rights practices in the state) are some examples of how some potential water problems (or problem areas) can be addressed.

(2) Use of Applicable Available Technologies

There are several technical tools that can assist in groundwater evaluations. Some of these include:

- well probes that can be used for determining accurate depths to the water table
- well drilling tools (and techniques) which allow for collection of water samples from different zones within a particular aquifer while an initial test hole is under construction
- remote-well sensing equipment that is designed to give continuous readouts, recording of static and pumping water levels, and other information from wells

(3) Determine Present and Future Demands on the Aquifer

A critical component of any groundwater monitoring plan should also ensure that the present use of the water does not endanger its use in the future. If water becomes scarce in the future, the costs of importing water into certain areas can be very prohibitive. It is therefore necessary to determine how much water is being withdrawn, how much is being recharged, and the effects that current and projected pumping activities have on the hydrologic balance of the aquifer(s).

Records on past water quantity use, present use, projected population and industrial growth, and estimates of per capita water use can help to predict present and projected demands on the aquifer(s). This basic information then can provide knowledge concerning how much water can be withdrawn safely from the aquifer without upsetting its hydrologic balance (e.g., Safe Yield Rate). Safe limits of water withdrawal from the aquifer can then be determined. If more water is being withdrawn than is being replenished, and a constant drop in the water table is occurring, then the aquifer is being mined. If this is the case, then a decision would have to be made concerning whether overdrafting should continue or efforts made to equilibrate the supply–demand function.

IV. SELECT COMPUTER MODELS FOR DATA SUMMARIZATION

A computerized data base information system should be included in any monitoring and protection plan, along with computers to facilitate analyses and summarization of any data generated from the previously mentioned steps (I–III). The data base system should also, as a minimum, be able to contain, analyze, and compare information concerning:

- hydrologic parameters of the aquifer
- delineation of recharge areas and zones
- historical, current, and projected water levels and water use
- known water quality parameters (biological, chemical, and radiological, etc.)

After groundwater monitoring related data is entered into a computerized data management system, the system should then be used to store incoming information.

(1) Use of Computer Models

Computer models can simulate characteristics of an aquifer and are able to predict how certain human activities will impact it (i.e., the drilling of an excessive number of high volume water wells in a vulnerable location; what a monitoring and protection program tries to prevent; Tung and Koltermann, 1985; Shirley, 1982; Olsthoorn, 1985; Monogham and Larson, 1985).

Some groundwater computer models are also able to translate mathematical results and interpret them back to physical conditions existing within an underground system. Any model, however, is meant to provide a predictive capacity that can be used to project water quality and quantity demands within an area. In this regard, many models are available, including some that can show the

needed proper spacing between wells and provide information concerning where wells should be constructed (Monogham and Larson, 1985; Tung and Koltermann, 1985). Artificial recharge rates within a given area can also be extrapolated via use of computer models.

The type of computer software that is appropriate for any groundwater monitoring network, depends on the computer hardware available to perform given analyses, and whether it be of the micro or large main-frame type. The type of questions asked, the number of personnel available to perform the work, budget limitations, and the data being collected from the groundwater systems being evaluated are also important.

The critical component of initially obtaining and maintaining reliable water related computerized information must also involve an ongoing Water Records Logging System (WRLS). This WRLS should provide a centralized information system data base concerning the status of any monitoring well in the area. The record-keeping system should also provide a standardized means or format for recording data concerning water levels, water quality parameters, water use, and water pumpage on a periodic basis (e.g., monthly, bimonthly, quarterly, or yearly sample results) from the water wells included in the monitoring network.

The contracted use of large main-frame computer facilities at local universities or the use of computers for records logging and analyses within larger municipalities should also be explored. Programs such as SAS (Statistical Analyses System, SAS Institute, 1985; SPSS-X, Statistical Package for the Social Sciences; Nie, 1983) can be modified to store groundwater monitoring data and summarize and conduct in-depth statistical analyses of the input parameters. Systems such as SAS and SPSS-X would also help in preparation of yearly or bi-yearly reports.

V. EVALUATE WATER LAW AND THE RESPONSIBILITIES OF MUNICIPALITIES AND GOVERNMENT

Water law maintains the rights of owners to utilize water that exists on or beneath their property (Trelease, 1974). The law relates to all individuals, including municipalities that use the same aquifer(s). In New Mexico, for example, the law provides that the surface and underground waters of the state belong to the public and are subject to appropriation for beneficial use. Such use is the basis, the measure, and the limit to the right to use water, with priority in time given to the better right. The underlying principle for water law in New Mexico is known as the Appropriative Doctrine, with water rights in the state being administered in accordance with provisions of the Constitution, the statutes, the terms of interstate water compacts, international treaties, and rules and regulations of the state engineer.

Local governments in most states, besides being regulated by their state water laws, also have the authority and responsibility to regulate urban expansion and provide public services, however, they might also want to seek direct involvement in all aspects of a resource such as water. They might want to get involved directly in groundwater monitoring and protection programs, including coordinating and being responsible for the entire system. State and federal efforts with groundwater monitoring and protection in this instance could still continue in the consultant capacity. However, in certain areas, if local beneficiaries are unwilling or unable to implement their own monitoring and protection programs, then state and federal governments should.

The federal U.S. EPA strategy is to strengthen state groundwater programs, including encouraging states to make use of existing grant programs to develop groundwater protection programs and strategies (U.S. EPA, 1984b). The agency also continues to encourage states to prepare or enhance their groundwater program development plans. Some states, such as New Mexico, had accomplished this prior to the new U.S. EPA strategy and have already prepared a program plan for the statewide monitoring of groundwater quality. The water quality monitoring plan for New Mexico includes sections on: conceptual framework for the statewide monitoring groundwater quality; identification of problem areas; priorities for data gathering; monitoring methods; and recommendations regarding development of a statewide groundwater monitoring system. Implementation of some of these phases for some areas of the state, however, is still lacking.

VI. PREPARE AND IMPLEMENT PLANS AND PROGRAMS

Steps (I–V) above should be considered and evaluated in developing any groundwater monitoring strategies. The steps presented are straightforward, and undoubtedly, as each component is designed and completed, unexpected situations specific to the particular aquifer(s) being evaluated will require additional work, time, and attention.

Development of groundwater monitoring and protection plans and programs are an evolving process. If plans and programs are successful, they will be a major contributor to the community's longevity because of a better understanding of the aquifer(s). A knowledge base of the aquifer(s) is important and necessary in terms of preventing degradation of water quality or implementing steps to remove contaminants from the aquifer(s), should they be discovered. Methods for abating or removing contamination from aquifer areas are limited and expensive, and the resulting contamination impacts may be health related, economic, or both.

Effective protection of groundwater against undue contamination requires well-integrated monitoring and protection plans and programs at the local, state, and federal levels. Eventually, substantial and comprehensive plans and programs that control and abate, as well as proposal to sta-

bilize underground water contamination, including identification of contaminated sources that can be corrected, will evolve.

Protection of aquifers must be stressed. A prevention-based approach along with groundwater cleanup is essential in terms of future policy decisions. These decisions must also involve the private sector. For it is in this area that problems are really put into the proper perspective and definite implementation of corrective and/or remedial action is implemented.

Overview of Groundwater Contamination—Summarization

Contamination of groundwater by organic and inorganic chemicals, radionuclides, and microorganisms has occurred in every state and is being detected with increasing frequency. For a long time, land surfaces and subsurface areas were considered safe and convenient depositories for many of society's wastes. However, only recently has the limited capacity of natural soil processes to change contaminants into harmless substances become widely recognized.

Detailed quantitative estimates of the extent and effects of groundwater contamination will never be available. The time, costs, and technical requirements to develop estimates on a nationwide basis is prohibitive. Information necessary for predicting future contamination problems (i.e., future uses of groundwater, potential sources, and types of contaminants) is also not known with certainty.

Contaminants observed in groundwater, particularly organic chemicals, are known to be associated with adverse health, social, environmental, and economic impacts. Although only a small portion of the nation's total groundwater resource is thought to be significantly contaminated on the overall scale, the potential effects of this contamination warrant national attention.

Public health concerns also arise because some groundwater contaminants are linked to cancers, liver and kidney damage, and damage to the central nervous system. Health concerns also arise because information is not available about adverse impacts of other individual contaminants, or of mixtures of contaminants occurring in groundwater. Uncertainties about human health impacts will continue to persist because these type of impacts are difficult to evaluate scientifically.

The health issues also become more complex, because some impacts are not observable until long after exposures have taken place. Social impacts related to health are hard to determine because they are often related to anxiety and sometimes fear about exposure to contaminants in groundwater. This fear can also occur unknowingly, because even if groundwater is contaminated, it oftentimes is odorless, colorless, and tasteless. Exposure may result gradually over many years.

In addition, environmental impacts involving groundwater contamination are not limited only to soil water movement, but must also include air and surface water areas because of the complex interrelationships water plays within ecosystems (e.g., groundwater is essential to streams, rivers, vegetation, fish, wildlife, etc.). The economic costs of detecting, correcting, and preventing groundwater contamination, therefore, can be high and will continue to be. Corrective and preventive actions will involve millions of dollars or more. On the other hand, economic losses that occur from impaired groundwater quality also relate significantly to decreases in agricultural and industrial productivity, lowered property values, costs for repair or replacement of damaged equipment, and the costs of developing usable alternative water supplies for domestic purposes.

Adverse environmental and economic impacts related to unnecessary groundwater contamination will increase. This is because contaminated groundwater occurs in industrialized and heavily populated areas. In addition, the continued human use of groundwater will be increasingly relied on as a principal source of water for many uses, although adverse health impacts will increase where contaminated groundwater is unknowingly consumed.

Although current information about groundwater contamination problems in the nation does not always describe actual situations, the information that has been gathered reflects the way in which investigations have been conducted, what contaminants have been searched, where they have been looked for, and where they have been detected. Because the majority of substances described as contaminants in groundwater are necessary for society, more widespread detection of contamination will continue to occur. Additional detection of contaminated groundwater areas will also be associated with increased efforts to monitor known problems, locate undetected problems, and monitor potential problems. Unfortunately, the costs and technical uncertainties associated with detection and correction activities effectively preclude the investigation and correction of known and/or suspected contamination problems. Consequently, prevention is central and essential to any long-term approach to groundwater quality protection. Choices involving detection, correction, and prevention, given limited funds and technical assistance, will depend on policy decisions regarding the extent groundwater resources are and should be protected.

Readings in μg/l and Depth in ft.

Table 1. Summarization of water quality data, Bernalillo County 1960–1976 (from: Rail, 1986; unpublished manuscript, Volume 1).

Variable	N	Mean	Standard Deviation	Minimum Value	Maximum Value	Std Error of Mean	Sum	Variance
Depth	967	112.36	148.67	15.0	1300.0	4.77	108653.00	22072.62
Na	413	93.86	219.28	2.8	3750.0	10.79	38767.36	48084.63
K	393	7.34	21.46	0.0	390.0	1.08	2902.70	460.74
Ca	422	77.02	82.47	0.0	707.0	4.01	32504.10	6801.69
Mg	411	15.48	24.09	0.0	325.0	1.18	6363.98	580.77
Fe	1264	0.75	3.85	0.0	54.7	0.10	952.32	14.80
Mn	1270	0.54	1.10	0.0	17.5	0.03	691.58	1.22
Cl	420	65.50	339.56	0.0	6300.0.	16.57	27513.50	115299.19
F	525	0.57	0.66	0.0	12.0	0.02	301.83	0.44
NO$_3$	667	35.44	172.83	0.0	2400.0	6.69	23640.77	29868.99
HCO$_3$	1011	244.01	151.87	0.0	2800.0	4.78	246696.43	23066.43
SO$_4$	1042	166.87	220.83	0.0	3971.0	6.84	173885.33	48767.29
Hardness	1225	276.16	222.78	0.0	2210.0	6.36	338303.75	49629.06
Cond.	1077	867.91	909.21	7.6	17100.00	27.71	934746.75	826677.90
pH	1025	7.97	0.32	6.2	9.88	0.00	8174.22	0.10

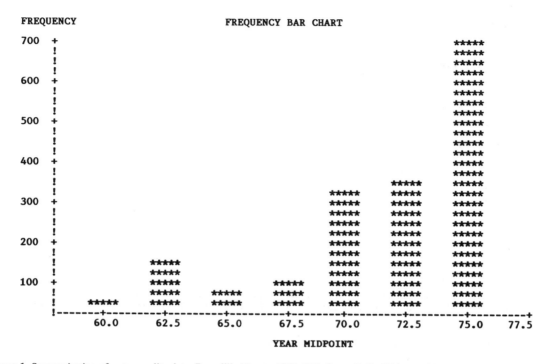

Figure 1. Summarization of water quality data, Bernalillo County 1960–1976 (from: Rail, 1986; unpublished manuscript, Volume 1).

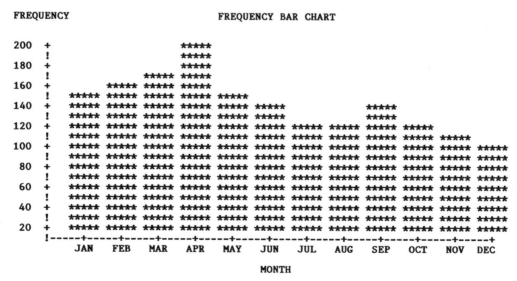

Figure 2. Summarization of water quality data, Bernalillo County 1960–1976 (from: Rail, 1986; unpublished manuscript, Volume 1).

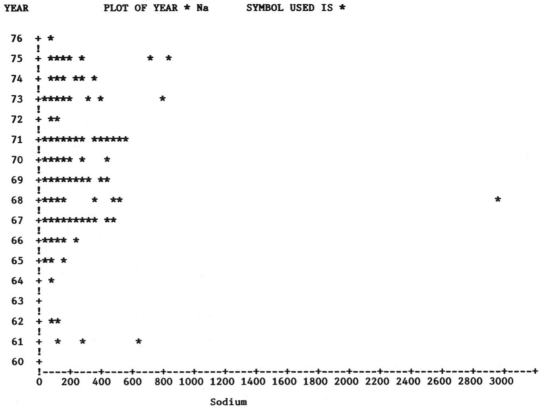

Figure 3. Summarization of water quality data, Bernalillo County 1960–1976 (from: Rail, 1986; unpublished manuscript, Volume 1).

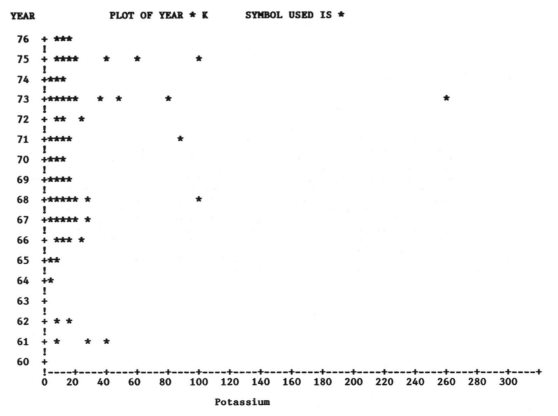

Figure 4. Summarization of water quality data, Bernalillo County 1960–1976 (from: Rail, 1986; unpublished manuscript, Volume 1).

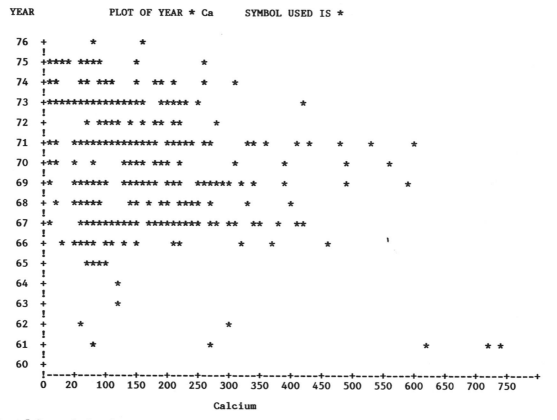

Figure 5. Summarization of water quality data, Bernalillo County 1960–1976 (from: Rail, 1986; unpublished manuscript, Volume 1).

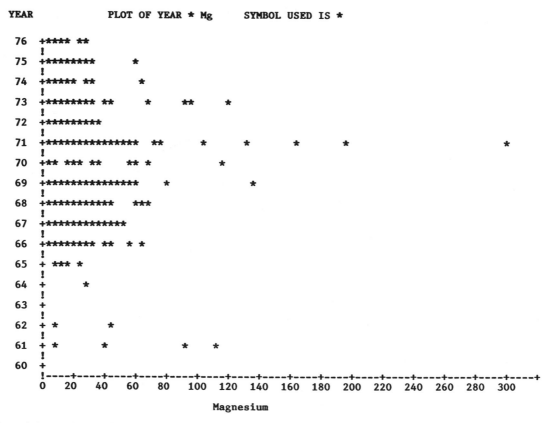

Figure 6. Summarization of water quality data, Bernalillo County 1960–1976 (from: Rail, 1986; unpublished manuscript, Volume 1).

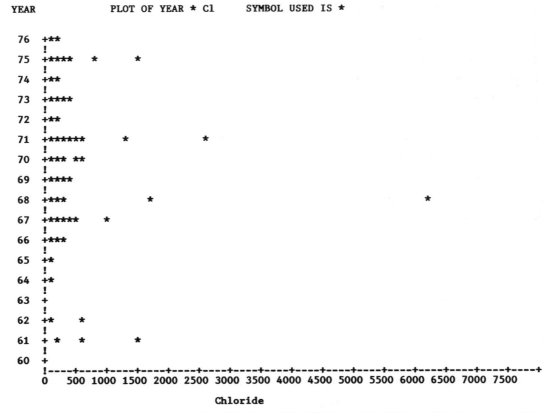

Figure 7. Summarization of water quality data, Bernalillo County 1960–1976 (from: Rail, 1986; unpublished manuscript, Volume 1).

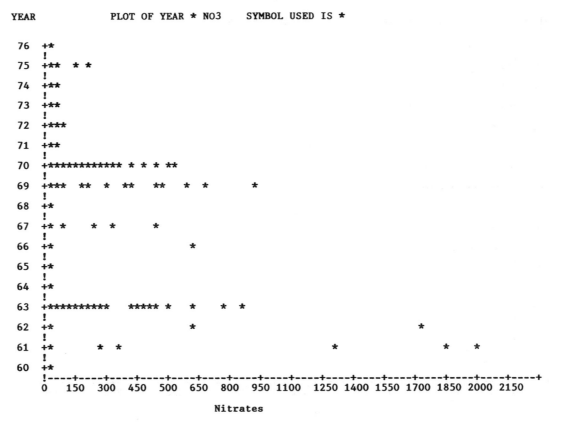

Figure 8. Summarization of water quality data, Bernalillo County 1960–1976 (from: Rail, 1986; unpublished manuscript, Volume 1).

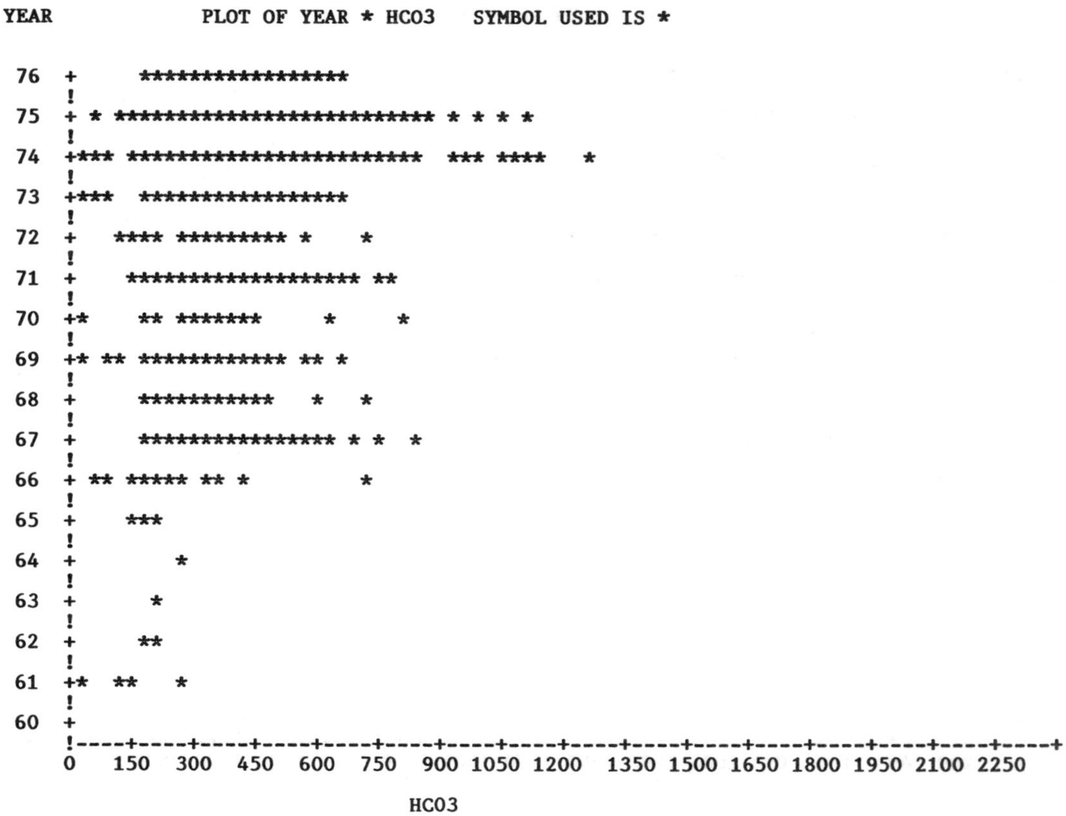

```
76 +     ****************
   !
75 + * ************************* * * * *
   !
74 +*** ************************ *** ****    *
   !
73 +*** *****************
   !
72 +   **** ********* *    *
   !
71 +    ****************** **
   !
70 +*    ** *******     *     *
   !
69 +* ** ************ ** *
   !
68 +      ***********    *    *
   !
67 +    *************** * *  *
   !
66 + ** ***** ** *         *
   !
65 +    ***
   !
64 +       *
   !
63 +      *
   !
62 +      **
   !
61 +* **  *
   !
60 +
   !----+----+----+----+----+----+----+----+----+----+----+----+----+----+----+----+
    0  150  300  450  600  750  900 1050 1200 1350 1500 1650 1800 1950 2100 2250

                              HCO3
```

Figure 9. Summarization of water quality data, Bernalillo County 1960–1976 (from: Rail, 1986; unpublished manuscript, Volume 1).

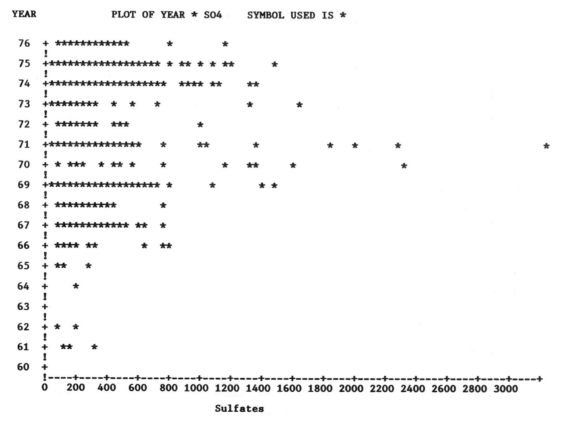

Figure 10. Summarization of water quality data, Bernalillo County 1960–1976 (from: Rail, 1986; unpublished manuscript, Volume 1).

111

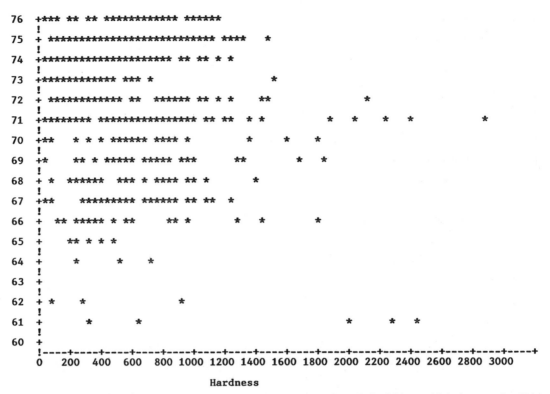

Figure 11. Summarization of water quality data, Bernalillo County 1960–1976 (from: Rail, 1986; unpublished manuscript, Volume 1).

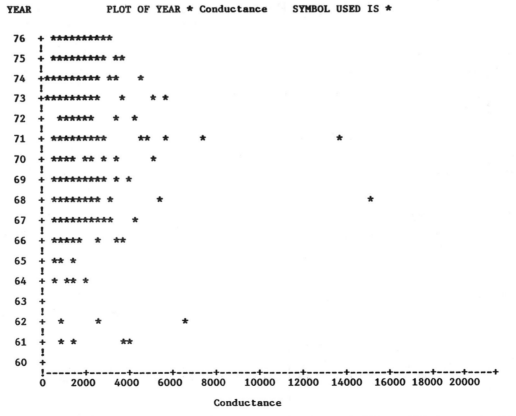

Figure 12. Summarization of water quality data, Bernalillo County 1960–1976 (from: Rail, 1986; unpublished manuscript, Volume 1).

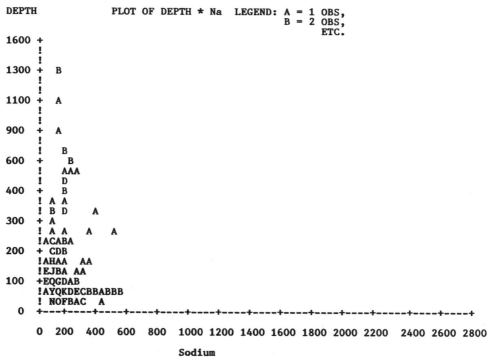

Figure 13. Summarization of water quality data, Bernalillo County 1960–1976 (from: Rail, 1986; unpublished manuscript, Volume 1).

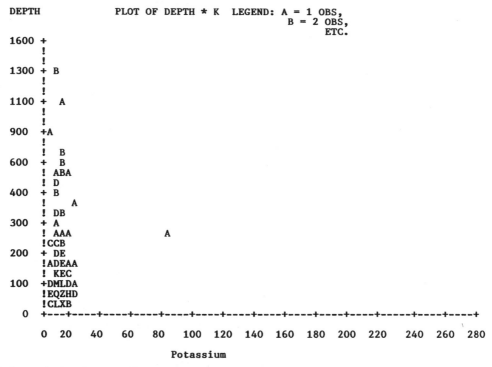

Figure 14. Summarization of water quality data, Bernalillo County 1960–1976 (from: Rail, 1986; unpublished manuscript, Volume 1).

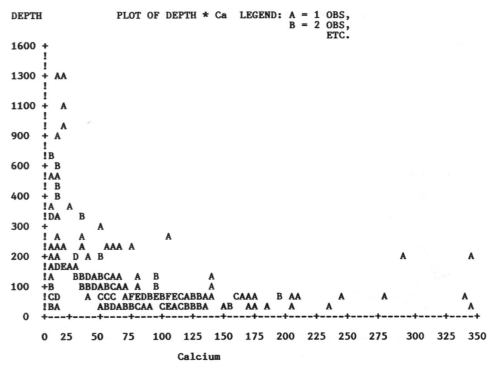

Figure 15. Summarization of water quality data, Bernalillo County 1960–1976 (from: Rail, 1986; unpublished manuscript, Volume 1).

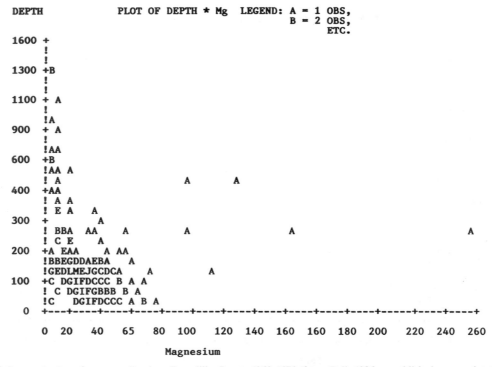

Figure 16. Summarization of water quality data, Bernalillo County 1960–1976 (from: Rail, 1986; unpublished manuscript, Volume 1).

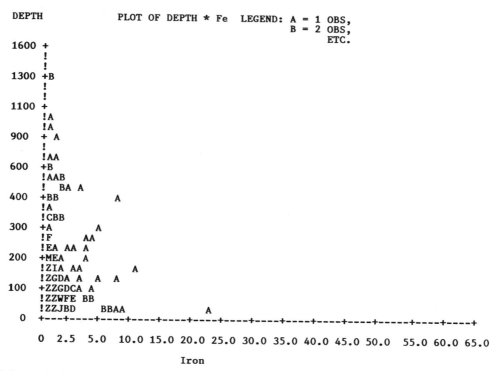

Figure 17. Summarization of water quality data, Bernalillo County 1960–1976 (from: Rail, 1986; unpublished manuscript, Volume 1).

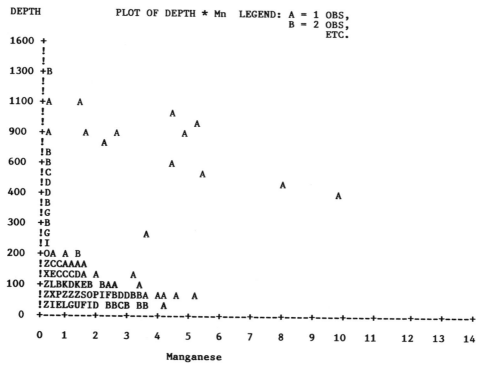

Figure 18. Summarization of water quality data, Bernalillo County 1960–1976 (from: Rail, 1986; unpublished manuscript, Volume 1).

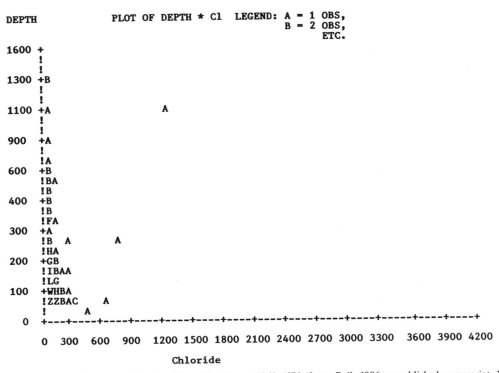

Figure 19. Summarization of water quality data, Bernalillo County 1960–1976 (from: Rail, 1986; unpublished manuscript, Volume 1).

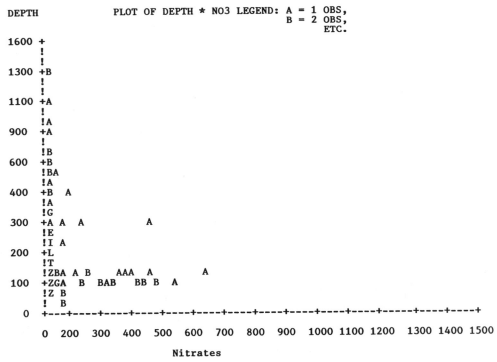

Figure 20. Summarization of water quality data, Bernalillo County 1960–1976 (from: Rail, 1986; unpublished manuscript, Volume 1).

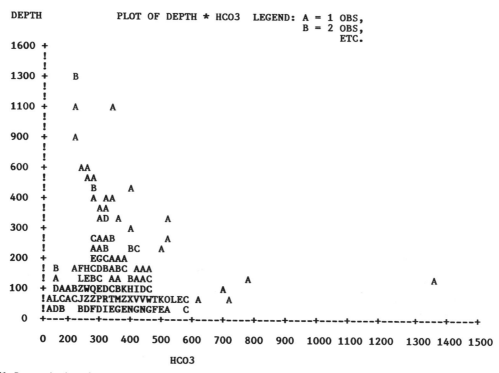

Figure 21. Summarization of water quality data, Bernalillo County 1960–1976 (from: Rail, 1986; unpublished manuscript, Volume 1).

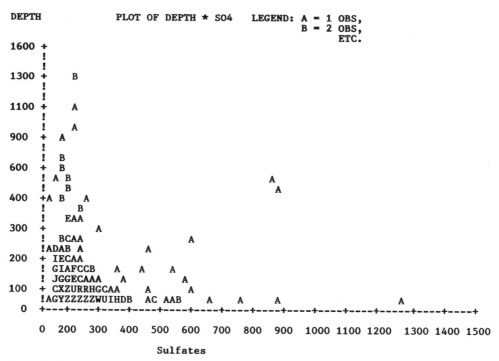

Figure 22. Summarization of water quality data, Bernalillo County 1960–1976 (from: Rail, 1986; unpublished manuscript, Volume 1).

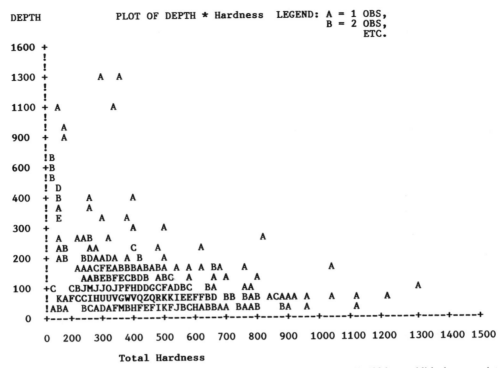

Figure 23. Summarization of water quality data, Bernalillo County 1960–1976 (from: Rail, 1986; unpublished manuscript, Volume 1).

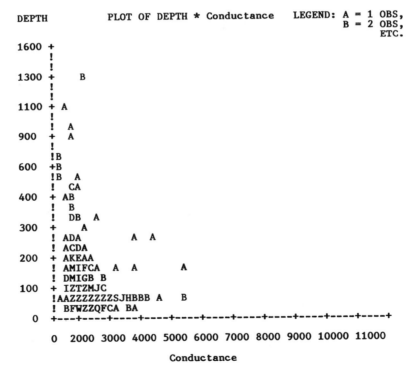

Figure 24. Summarization of water quality data, Bernalillo County 1960–1976 (from: Rail, 1986; unpublished manuscript, Volume 1).

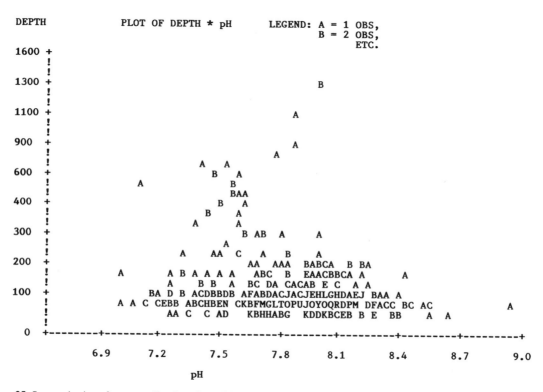

Figure 25. Summarization of water quality data, Bernalillo County 1960–1976 (from: Rail, 1986; unpublished manuscript, Volume 1).

Checklist of Concerns to Prevent
Groundwater Contamination Problems

(Adapted from: Water-Resource-Management Strategy, City of Albuquerque, July, 1987).

I. Demand-Forecast Element
Principal Tasks:
- collect population data and forecasts
- collect water-use data and projections
- collect City of Albuquerque future-development plans, including:
  - the Comprehensive Plan
  - area plans approved during the last five years
  - recently accepted and proposed water master plans
  - recently accepted transportation master plan
  - recently accepted sewer master plan
  - recently accepted drainage-area master plans
- collect all county and regional plans
- systematically compare the plans with the existing system and determine changes in the water distribution system that each plan implies
- predict the probable service area and identify the probable subareas within it
- estimate the population growth, number of water users, their demand for water, and the peak demands during the years 2000, 2025, and 2050
- prepare a detailed report that contains the population and water used data, describe the methods used, and tabulate the size subareas

II. Groundwater-Use Element
Principal Tasks:
- compile construction history, equipment history, and related data of each well
- develop a formula to define for any well the depth to water
- using the U.S. Geological Survey's model, predict:
  - when each well reaches its economic maximum depth to water
  - effects of other nearby wells on the Department's wells

- the effects of adding wells at the locations to supply water into the future
- determine the optimum distance between wells
- change in the flow the Rio Grande caused by pumping through 2050
- estimate the population the groundwater resource will serve
- prepare a report that describes the simulations, states the results of each simulation, and discusses the implications of the simulation on the Department's capacity to deliver water

III. Geologic Element
Principal Tasks:
- prepare maps of the drainage basin that define its position
- prepare a stratigraphic column of the Albuquerque basin
- for geomorphic features prepare maps of progressive larger portions
- collect published and open file geologic reports and maps
- collect well and test-hole logs and bore-hole geophysical surveys of test holes
- drill, log, sample, and run bore-hole geophysical surveys of test holes
- measure hydraulic properties of rocks using wells or piezometers
- prepare descriptive logs of cuttings from (a) existing wells and test holes and (b) new wells and test (pilot) holes
- develop data-management schemata
- prepare surface geologic maps of progressively larger portions of the basin
- prepare hydraulic-properties maps for each strati-

(continued)

Checklist of concerns that relate to prevention of groundwater contamination problems
(Adapted from: Water-Resource-Management Strategy, City of Albuquerque, July, 1987) *(continued)*.

III. Geological Element
Principal Tasks: (continued)
graphic unit for increasingly larger portions of the Albuquerque basin

IV. Pollution Element
Principal Tasks:
- develop definitions and classifications
- prepare maps of three increasingly larger portions of the basin that show landfills, dumps, septic-tank concentrations, sewers, buried tanks, etc.
- within the city investigate sites to quantify rates and areas of pollution
- outside the city investigate sites to quantify rates and areas of pollution
- prepare maps that show concentrations of pollutants or pollution indicators
- develop a pollution prevention program
- develop and initiate remedial measures
- develop monitoring programs to detect pollutants before they become a hazard
- develop contingency plans by hydrologic setting to clean up new pollution
- develop data-management schemata

V. Monitoring Element
Principal Tasks:
- Prepare a catalog that:
  Lists:
  - the parameters currently monitored or measured
  - the parameters that have been used by numerical models elsewhere
  Tells:
  - who makes the measurements
  - how they make the measurements
  - how the measurements are made elsewhere
- Prepare a list that tells:
  - which parameters the city should monitor
  - how the measurements should be made
  - by whom, where, how often, and the form of the data report
- Institute a data-collection program and develop a data-management schemata

VI. Data-Management Element
Principal Tasks:
- develop a catalog of data that the system must manage
- develop a generic list of tasks that the system must perform
- inventory users and possible users to determine specific tasks performed
- develop a program cataloging and storing procedure for hard copy
- amass hard-copy books, reports, records, documents, logs, etc.
- provide for the addition of new hard copy as it becomes available
- develop a program for making hard copy available to users
- develop procedures to handle data from either automated or non-automated data collection or generating programs
- inventory the computer software/hardware that might serve
- develop computer programs to handle automated data collected by monitoring systems
- design and install the computer system

Ahmad, M. U. "A Hydrological Approach to Control Acid Mine Pollution for Lake Hope," *Ground Water,* 8(6):19–24 (1970).

Aieta, E. M. et al. "Radionuclides in Drinking Water: An Overview," *Journal American Water Works Association,* 79(4):144–153 (1987).

Aley, T. J. et al. "Groundwater Contamination and Sinkhole Collapse Induced by Leaky Impoundments in Soluble Rock Terrain," *Eng. Geol. Series No. 5,* Rolla, Missouri:Missouri Geological Survey and Water Resources, 32 pp. (1972).

Allen, J. M. and E. E. Geldreich. "Bacteriological Criteria for Groundwater Quality," *Ground Water,* 13:1–7 (1975).

American Gas Association, Inc. Survey of Gas Storage Facilities in United States and Canada. New York (1967).

American Gas Association, Inc. Survey of Gas Storage Facilities in United States and Canada. New York (1971).

American Petroleum Institute. "The Migration of Petroleum Products in Soil and Ground Water." Wash., D.C.:American Petroleum Inst. Publ. No. 4149 (1972).

American Petroleum Institute. "Guide to Ground Water Standards of the United States." Wash., D.C.:Am. Petro. Inst. (1983a).

American Petroleum Institute. "Underground Leak Laws." Wash., D.C. (1983b).

American Public Health Association. *Standard Methods for the Examination of Water and Wastewater.* 12th edition, Washington, D.C. (1965).

American Public Health Association. *Standard Methods for the Examination of Water and Wastewater.* 13th edition, Washington, D.C. (1975).

American Public Health Association. *Standard Methods for the Examination of Water and Wastewater.* Washington, D.C. (1986).

American Society for Testing Materials. *Annual Book of ASTM Standards.* pt. 23, "Water and Atmospheric Analyses," Philadelphia, PA (1970).

American Water Works Association. *Water Quality and Treatment.* 2nd edition (1950).

Anderson, M. D. and C. J. Bowser. "The Role of Groundwater in Delaying Lake Acidification," *Water Resources Research,* 22(7):1101–1108 (July 1986).

Anonymous. "Action of Pure Water on Cement Mortar Briquets," *Journal American Water Works,* 20:585 (1938).

Anonymous. "Standards for Water Quality and Factors Affecting Quality," *Public Health Engineering Abstracts,* 27:86 (1947).

Anonymous. *Water Quality and Treatment.* 2nd edition. American Water Works Association (1950).

Anonymous. *Handbook of Chemistry and Physics.* 32nd edition. Chemical Rubber Co. (1950b).

Anonymous. "California Acts to Cut Water Pollution," *Oil and Gas Journal,* 55(28):66–67 (1957).

Anonymous. "Manganese in Reservoirs," *Industrial Wastes,* 3:3, 9A (1958).

Anonymous. *The Merck Index of Chemicals and Drugs.* 7th edition (1960).

Anonymous. "Deep Injection Wells," *Water Well Journal,* 22(8):12–13 (1968a).

Anonymous. "Deep-Well Injection Is Effective for Waste Disposal," *Environmental Science and Technology,* 2(6):406–410 (1968b).

Anonymous. "Pollution-Free Sewage Disposal," *Ground Water Age,* 7(10):21–22, 27–28 (1973).

Anonymous. "On-Site Treatment for Low-Density Areas," *American City and County* (April 1980).

Anonymous. "Rocky Mountain Arsenal: Landmark Case of Groundwater Polluted by Organic Chemicals, and Being Cleaned Up," *Civil Engineering,* 51:68–71 (1981).

Archey and Mawson. "Municipal Watershed Management—A Unique Opportunity in Massachusetts," *Journal NEWWA* (June 1984).

Bachmat, Y. et al. (eds.). *Groundwater Management. The Use of Numerical Models.* Water Resource Monographs. 5. Washington, D.C.:AGU (1980).

Bacon, M. J. and W. A. Oleckno. "Groundwater Contamination: A National Problem with Implications for State and Local Environmental Health Personnel," *Journal of Environmental Health,* 48(3):116–121 (1985).

Baffa, J. J. and N. J. Bartilucci. "Wastewater Reclammation by Groundwater Recharge on Long Island, New York," *Geological Survey Research 1966,* 39(3):Part 1, 431–445 (1967).

Baker, E. T., Jr. and J. Rawson. "Ground-Water Pollution in the Vicinity of Toledo Bend Reservoir, Texas." Progress Report. U.S. Geological Survey Open-File Report (August 1972).

Barbash, J. and P. Roberts. "Volatile Organic Chemical Contamination of Groundwater Resources in the U.S.," *Journal WPCF,* 58:5 (1986).

Barr, A. J. et al. *SAS Users Guide.* SAS Inst., Inc., Raleigh, N.C.:Spark Press (1979).

Barraclough, J. T. "Waste Injection into a Deep Limestone in Northwestern Florida." *Ground Water,* 4(1):22–24 (1966).

Barrett, P. H., "Relationships between Alkalinity and Absorption and Regeneration of Added Phosphorus in Fertilized Trout Lakes," *Trans. Amer. Fish. Soc.,* 82:78 (1952).

Barrs, J. K. "Travel of Pollution and Purification En Route in Sandy Soils," *Bulletin of the World Health Organization,* 16:727–747 (1957).

Bates, R. G. *Determination of pH—Theory and Practice.* New York:John Wiley and Sons, 435 pp. (1964).

Bean, E. H. "Progress Report on Water Quality Criteria," *American Water Works Association Journal,* 54:1313–1331 (1962).

Belter, W. G. "Waste Management Activities in the Atomic Energy Commission," *Ground Water,* 1(1):11–15 (1963).

Bendixon, T. W. et al. "Ridge and Furrow Liquid Waste Disposal in a Northern Latitude," *Journal Sanitary Engineering Division*, American Society of Civil Engineers, 94(SA 1):147–157 (1968).

Bennett, R. R. and R. R. Meyer. "Geology and Groundwater Resources of the Baltimore Area," Maryland Dept. of Geology, Mines, and Water Resources Bulletin 4 (1952).

Bergstrom, R. E. "Feasibility Criteria for Subsurface Waste Disposal in Illinois," *Ground Water*, 6(5):5–9 (1968).

Bergstrom, R. E. "Hydrogeological Studies Are Key to Safety in Waste Management Programs," *Water and Sewage Works*, 116(4):149–155 (1969).

Bernhart, A. P. "Protection of Water-Supply Wells from Contamination by Wastewater," *Ground Water*, 11(3):9–15 (1973).

Biggar, J. W. and R. B. Corey. "Agricultural Drainage and Eutrophication," *Eutrophication: Causes, Consequences, Correctives*. Washington, D.C.:National Academy of Sciences, pp. 404–445 (1969).

Bitton G. "Viruses in Drinking Water," *Environmental Science and Technology*, 20(3) (1986).

Bloss, F. D. and R. L. Steiner. "Biogeochemical Prospecting for Manganese in Northeast Tennessee," *Geological Society of American Bulletin*, 71:1053–1066 (1960).

Boettner, E. A. and F. I. Grunder. "Water Analysis by Atomic Absorption and Flameless Emission Spectroscopy," in R. A. Baker (ed.) *Trace Inorganics in Water*, pp. 236–246, Adv. Chem. Ser. No. 73, American Chemical Society, Washington, D.C. (1968).

Bonde, E. K. and P. Urone. "Plant Toxicants in Underground Water in Adams County, Colorado," *Soil Science*, 93(5):353–356 (1962).

Born, S. M. and C. A. Stephenson. "Hydrogeologic Considerations in Liquid Waste Disposal," *Journal Soil and Water Conservation*, 24(2):52–55 (1969).

Boucher, R. R. and G. F. Lee, "Adsorption of Lindane and Dieldrin Pesticides on Unconsolidated Aquifer Sands," *Environmental Science and Technology*, 6(6):538–543 (1972).

Bouwer, H. et al. "Renovating Secondary Sewage by Groundwater Recharge with Infiltration Basins," U.S. Environmental Protection Agency, Water Pollution Control Research Series 16070-DRV, 102 pp. (1972).

Bouwer, H. "Returning Wastes to the Land, a New Role for Agriculture," *Journal Soil and Water Conservation*, 23(5):164–169 (1968).

Bouwer, H. "Water-Quality Improvement by Ground-Water Recharge," Agricultural Research Service, Report 41–147 pp. 23–27 (1969).

Bouwer, H. "Groundwater Recharge Design for Renovating Waste Water," *Journal of Sanitary Engineering Divison*, American Society of Civil Engineers, 96(SA 1):59–74, Paper 7096 (1970).

Bouwer, H. "Effect of Irrigated Agriculture on Groundwater," *Journal of Irrigation and Drainage Engineering*, 113(1):4–15 (1987).

Bowman, I. "Well Drilling Methods," U.S. Geological Survey Water-Supply Paper 257 (1911).

Brashears, M. C. Jr. "Ground-water Temperatures on Long Island, New York as Affected by Recharge of Warm Water," *Economic Geology*, 36:811–828 (1941).

Bromehead, C. E. N. "The Early History of Water-Supply," *The Geographical Journal*, 99:142–151; 183–196 (1942).

Brown, R. F. and D. C. Signor. "Groundwater Recharge," *Water Resources Bulletin*, 8(1):132–149 (1972).

Brown, R. E. et al. "Geological and Hydrological Aspects of the Disposal of Liquid Radioactive Wastes," *Proceedings of Seminar on Sanitary Engineering Aspects of the Atomic Energy Commission Div. of Technical Information*, Report TID-7517, Part 16, pp. 413–425 (1956).

Browning, E. *Toxicity of Industrial Metals*. London, England:Butterworths (Dec. 1961).

Burke, R. G. "Texas Toughens Anti-Pollution Line," *Oil and Gas Journal*, 64(1):47–48 (1966).

Burmaster, D. E. and R. H. Harris. "Groundwater Contamination: An Emerging Threat," *Technology Review*, 85(5):50–62 (1982).

Butler, H. C. *The Story of Athens*, Century Co., pp. 74–75 (1902).

Butler, G. C. *Principles of Ecotoxicology*. New York:Wiley (1978).

Butler, M. A. "Irrigation in Persia by Kanats," *Civil Engineering*, 3(2):69–73 (1933).

California Department of Water Resources. "Ground Water Basin Protection Projects; Oxnard Basin Experimental Extraction-Type Barrier," Bulletin 147-6, 157 pp. (1970).

Campbell, M. D. and J. H. Lehr. *Water Well Technology*. New York:McGraw-Hill, 681 pp. (1973).

Canter, L. W. and R. C. Knox. *Septic Tank System Effects on Ground Water Quality*. Chelsea, Michigan:Lewis Publishers, Inc. (1985).

Canter, L. W. and R. C. Knox. *Ground Water Pollution Control*. Chelsea, Michigan:Lewis Publishers, Inc., 525 pp. (1986).

Carroll, D. "Rainwater as a Chemical Agent of Geologic Processes—A Review," U.S. Geological Survey Water-Supply Paper 1535-G, 18 pp. (1962).

Caswell, C. A. "Underground Waste Disposal," *Environmental Science and Technology*, 4(8):655–666 (1970).

Chaiken, E. I. et al. "Muskeggon Sprays Sewage Effluents on Land," *Civil Engineering*, 43(5):49–53 (1973).

Champlin, J. B. F. and G. G. Eichholz. "The Movement of Radioactive Sodium and Ruthenium through a Simulated Aquifer," *Water Resources Research*, 4(1):147–158 (1968).

Chebotarev, I. "Metamorphism of Natural Waters on the Crust of Weathering," *Geochim. et Cosmochim. Acta.*, 8:22–48; 137–170; 198–212 (1955).

Clark, G. "Water in Antiquity," *Antiquity*, XVIII(69) (Mar. 1944).

Clarke, F. W. and H. S. Washington. "The Composition of the Earth's Crust," U.S. Geological Survey, Professional Paper 127, 117 pp. (1924).

Clarke, F. W. *The Data of Geochemistry*. 5th edition, U.S. Geological Survey Bulletin No. 770, 841 pp. (1924).

Cleary, E. J. and D. L. Warner. "Some Considerations in Underground Waste Water Disposal," *Journal American Water Works Association*, 62(8):489–498 (1970).

Cohen, P. and C. N. Durfor. "Design and Construction of a Unique Injection Well on Long Island, New York," Geological Survey Research 1966, U.S. Geological Survey Professional Paper 550-D, pp. D253–D257 (1966).

Cohen, J. M. "Taste Threshold Concentrations of Metals in Drinking Water," *Journal American Water Works Association*, 52:660 (1960).

Collins, W. D. "Water for Industrial Purposes," *American City* (July, August, September 1937).

Collins, A. G. "Oil and Gas Wells—Potential Polluters of the Environment?" *Journal Water Pollution Control Federation*, 43(12):2383–2393 (1971).

Committee on Environmental Affairs. "The Migration of Petroleum Products in Soil and Ground Water—Principles and Countermeasures," Publication No. 4149, American Petroleum Institue, Washington, D.C., 36 pp. (1972).

Committee on Ground Water, *Ground Water Management*, Manual No. 40, American Society of Civil Engineers, New York, 216 pp. (1972).

Cogniglio, W. "Criteria and Standards Division Briefing on Occurrence/Exposure to Volatile Organic Chemicals," Office of Drinking Water, U.S. Environmental Protection Agency, Washginton, D.C. (1982).

Connelley, E. J., Jr. "Removal of Iron and Manganese," *Journal American Water Works Associaton*, 50:697 (1958).

Conservation Foundation. *Groundwater, Saving the Unseen Resource*. Proposed conclusions and recommendations. The National Groundwater Policy forum. Sponsored by the Conservation foundation in cooperation with the National Governor's Association. Washington, D.C. (1985).

Cooper, H. H., Jr. et al. "Sea Water in Coastal Aquifers," U.S. Geological Survey Water-Supply Paper 1613-C, 84 pp. (1964).

Cornaby, B. W. et al. "Application of Environmental Risk Techniques to Uncontrolled Hazardous Waste Sites," *Proc. of the Nat. Conf. on Management of Uncontrolled Hazardous Waste Sites, 1982*, Hazardous Materials Control Research Inst., Silver Spring, Maryland, pp. 390–395 (1982).

Cosby, B. J. et al. *Environmental Science and Technology*, 19:1144–1149 (1985).

Cosgrove, J. F. and D. J. Brocco. "Determination of Minor Metallic Elements in the Water Environment," in Ciacio (ed.), *Water and Water Pollution Handbook, Vol. 4*, pp. 1315–1356, New York:M. Dekker Co. (1973).

Cothern, C. R. "Estimating the Health Risks of Radon in Drinking Water," *Journal American Water Works Association*, 79(4):153 (1987).

Coulborn, R. "The Ancient River Valley Civilizations," in *New Perspectives in World History*, S. H. Engle, ed., Washington (1964).

Council of Environmental Quality. *The Eleventh Report of the Council of Environmental Quality*. Washington, D.C.:Government Printing Office, 497 pp. (1980).

Council of Environmental Quality. *Contamination of Ground Water by Toxic Organic Chemicals*. Supt. of Documents, Washington, D.C.:U.S. Government Printing Office (1981).

Craun, G. F. "Outbreaks of Waterborne Disease in the United States: 1971–1978," *Journal American Water Works Association*, 73:360–369 (1981).

Crosby, J. W., III et al. "Migration of Pollutants in a Glacial Outwash Environment, 2," *Water Resources Research*, 7(1):204–208 (1971).

Curran, C. M. and R. B. Norton. *EOS Trans. Am. Geophys. Union*, 66:823, Abstract, A22A-09 (1985).

Dair, F. R. "Seepage and Seepage Control Problems in Sanitary Landfills," Agricultural Research Service, ARS 41–147, U.S. Dept. Ag., pp. 14–16 (1969).

Davids, H. W. and M. Lieber. "Underground Water Contamination by Chromium Wastes," *Water and Sewage Works*, 98:528–534 (1951).

Davis, S. N. and J. M. DeWiest. *Hydrogeology*. 2nd ed., New York, London, Sydney:Wiley, 436 pp. (1967).

Dean, B. T. "The Design and Operation of a Deep-Well Disposal System," *Journal Water Pollution Control Federation*, 37(2):245–254 (1965).

Department of Transportation. Title 49, "Transportation," *Federal Register, Vol. 34*, No. 191, Washington, D.C. (1969).

Department of Transportation. Part 195, "Transportation of Liquids by Pipeline," *Federal Register, Vol. 35*, No. 218, Washington, D.C. (1970).

Department of Transportation. Title 49, "Transportation," *Federal Register, Vol. 35*, No. 218, Washington, D.C. (1970a).

Deutsch, M. "Phenol Contamination of an Artesian Glacial-Drift Aquifer at Alma, Michigan, U.S.A.," *Soc. for Water Treatment and Examination Proc. 11, Pt. 2*, pp. 94–100 (1962).

Deutsch, M. "Groundwater Contamination and Legal Controls in Michigan," *U.S. Geological Survey Water-Supply 1691*, 78 pp. (1963).

DeLaguna, W. and J. O. Blomeke. "The Disposal of Power Reactor Waste into Deep Wells," Atomic Energy Commission Report ORNL-CF-57-6-23, U.S. Atomic Energy Commission Office of Technical Information, 48 pp. (1957).

DeWalle, F. B. and R. M. Schaff. "Groundwater Pollution by Septic Tank Drain Fields," *Journal Environmental Engineering Division*, American Soc. Civil Engr., 106(EE3):631–648 (1980).

Domencio, P. A. *Concepts and Models in Groundwater Hydrology*, New York:McGraw Hill, 405 pp. (1972).

Donaldson, E. C. "Subsurface Disposal of Industrial Wastes in the United States," U.S. Bureau of Mines Information Circular 8212, 34 pp. (1964).

Doudoroff, P. and M. Katz. "Critical Review of Literature on the Toxicity of Industrial Wastes and Their Components to Fish. II. The Metals as Salts," *Sewage and Industrial Wastes*, 25(7):802 (1953).

Dregne, H. E. et al. "Movement of 2,4-D in Soils," Western Regional Research Project Progress Report. New Mexico Agricultural Experiment Station, University Park, 35 pp. (1969).

Driscoll, C. T. and R. M. Newton. *Environmental Science and Technology*, 19:1018–24 (1985).

Durfor, C. N. and E. Becker. "Public Water Supplies of the 100 Largest Cities in the United States," U.S. Geological Water-Supply Paper 1812, 364 pp. (1964).

Eddy, G. E. "The Effectiveness of Michigan's Oil and Gas Conservation Law in Preventing Pollution of the State's Ground Waters," *Ground Water*, 3(2):35–36 (1965).

Eikum, A. S. and B. Paulsrud. "Treatment of Septage-Scandanavian Practice," *Water Science and Technology*, 18:63–70 (1986).

Ellis, J. B. et al. "Hydrological Controls of Pollutant Removal from Highway Surfaces," *Water Resources*, 20(5):589–595 (1986).

Elwell, W. T. and J. A. Gidley. *Atomic Absorption Spectrophotometry*. 2nd. ed., New York:Pergamon Press (1966).

Emrich, G. H. and R. A. Landon. "Investigation of the Effects of Sanitary Landfills in Coal Strip Mines on Ground Water Quality," Pennsylvania Bureau of Water Quality Management, Publication No. 30, 37 pp. (1971).

Emrich, G. H. and G. L. Merrit. "Effects of Mine Drainage on Ground Water," *Ground Water*, 7(3):27–32 (1969).

Enock, C. R. *The Secret of the Pacific*. New York, London:Charles Scribner's Sons (1912).

Evans, D. M. and A. Bradford. "Under the Rug," *Environment*, 11(8):3–13 (1969).

Evans, D. M. "The Denver Area Earthquakes and the Rocky Mountain Arsenal Disposal Well," *Mountain Geologist*, 3(1):23–26 (1966).

Evans, R. "Industrial Wastes and Water Supplies," *Journal American Water Works Association*, 57(4):625–628 (1965).

Eye, J. D. "Aqueous Transport of Dieldrin Residues in Soils," *Journal Water Pollution Control Federation*, 40(6):538–543 (1968).

Farrara, R. A. et al. *Groundwater Contamination from Hazardous Wastes*. Princeton University Water Resources Program, New Jersey:Prentice-Hall, Inc. (1984).

Federal Water Pollution Control Administration. "Water Pollution Aspects of Urban Runoff," Bull. WP 20-15, Federal Water Pollution Control Ad., 272 pp. (1969).

Fleischer, M. "Recent Estimates of the Abundance of Elements in the Earth's Crust," U.S. Geological Survey Circular 285, 7 pp. (1953).

Fleischer, M. "The Abundance and Distribution of the Chemical Elements in the Earth's Crust," *Journal Chemical Education*, 31:446–455 (1954).

Foose, R. M. "Sinkhole Formation by Groundwater Withdrawal, Far West Rand, South Africa," *Science*, 157(3792):1045–1048 (1967).

Fraser, D. *The Marches of Hindustan*. London:Blackwood and Sons (1907).

Freeze, R. A. and J. A. Cherry. *Groundwater*. Englewood Cliffs, New Jersey:Prentice Hall, 604 pp. (1979).

Fryberger, J. S. "Rehabilitation of a Brine-Polluted Aquifer," U.S. Environmental Protection Agency, Environmental Protection Technology Series EPA-R2-72-014, Washington, D.C., 61 pp. (1972).

Fuhriman, D. K. and J. R. Barton. "Ground Water Pollution in Arizona, California, Nevada, and Utah," U.S. Environmental Protection Agency, Water Pollution Control Research Series 16061, 249 pp. (1971).

Gaffney, J. S. et al. "Beyond Acid Rain," *Environmental Science and Technology*, 21(6):519–523 (1987).

Galbraith, H. H. et al. "Migration and Leaching of Metals from Old Mine Tailings Deposits," *Ground Water*, 10(3):17–26 (1972).

Gambell, A. W. and D. W. Fisher. "Chemical Composition of Rainfall in Eastern North Carolina and Southeastern Virginia," U.S. Geological Survey Water-Supply Paper 1535-K, 41 pp. (1966).

Garcia-Bengochea, J. L. and R. O. Vernon. "Deep Well Disposal of Waste Waters in Saline Aquifers of South Florida," *Water Resources Research*, 6(5):1464–1470 (1970).

Garrels, R. M. and C. L. Christ. *Solutions, Minerals, and Equilibria*. New York:Harper and Row, 450 pp. (1964).

Gass, T. E. "Ground Water in the News," *Ground Water*, 23:148–149 (1985).

Gerba, C. P. et al. "Viruses in Water: The Problem, Some Solutions," *Environmental Science and Technology*, 9:1122–1126 (1975).

Gibson, U. P. and R. D. Singer. *Water Well Manual*. Berkeley, California: Premier Press, 156 pp. (1971).

Gilbertson, C. B. et al. "Runoff, Solid Wastes, and Nitrate Movement on Beef Feedlots," *Journal Water Pollution Control Federation, Part 1*, 43(3):483–493 (1971).

Gillham, R. W. and L. R. Webber. "Groundwater Contamination," *Water and Pollution Control*, 106(5):54–57 (1968).

Gillham, R. W. and L. R. Webber. "Nitrogen Contamination of Groundwater by Barnyard Leachates," *Journal Water Pollution Control Federation*, 41(10):1752–1762 (1969).

Goldschmitt, V. M. *Geochemistry.* Alex, Muir, ed. Oxford, England:Clarendon Press, 730 pp. (1954).

Golson, J., ed. *Polynesian Navigation.* A symposium on Andres Sharp's theory of accidental voyages. The Polynesian Society, Wellington, N.Z. (1963).

Golueke, C. G. and P. H. McGauhey. "Comprehensive Studies of Solid Waste Management," First Annual Report, SERL Report No. 67-7. Sanitary Engineering Research Laboratory, University of California, Berkeley (1967).

Goolsby, D. A. "Hydrogeochemical Effects of Injecting Wastes into a Limestone Aquifer Near Pensacola, Florida," *Ground Water,* 9(1):13–19 (1971).

Gottleib, M. S. et al. "Drinking Water and Cancer in Louisiana: A Retrospective Mortality Study," *Am. J. Epid.,* 116:652 (1985).

Gregg, D. O., "Protective Pumping to Reduce Aquifer Pollution, Glynn County, Georgia," *Ground Water,* 9(5):21–29 (1971).

Grieder, T. *Origins of Pre-Columbian Art.* Austin, Texas:University of Texas Press (1982).

Grier, N. "Backyard Poison: A Look at Urban Pesticide Use," *Northwest Coalition Alternatives to Pesticide News,* 3(1):50–51 (1981–82).

Griffin, A. E. "Problems Caused by Manganese in Water Supplies," *Journal American Water Works Association,* 50:1386 (1958).

Griffin, A. E. "Significance and Removal of Manganese in Water Supplies," *Journal American Water Works Association,* 52:1326 (1960).

Grubb, H. F. "Effects of a Concentrated Acid on Water Chemistry and Water Use in a Pleistocene Outwash Aquifer," *Ground Water,* 8(5):4–7 (1970).

Guttman-Bass, N. and B. Fattal. "Analysis of Tap Water for Viruses: Results of a Survey," *Water Science and Technology,* 17(10) (1984).

Hackbarth, D. A. "Field Study of Subsurface Spent Sulfite Liquor Movement Using Earth Resistivity Measurements," *Ground Water,* 9(3):11–16 (1971).

Hackett, J. E. "Water Resources and the Urban Environment," *Ground Water,* 7(2):11–14 (1969).

Hain, K. E. and R. T. O'Brian. "The Survival of Enteric Viruses in Septic Tanks and Septic Tank Drain Fields," Water Resources Research Institute, New Mexico State University, Las Cruces, Rep. No. 108 (1979).

Hale, W. E. et al. "Characteristics of the Water Supply in New Mexico," Tech. Rep. 31, New Mexico State Engr. and U.S. Geological Survey, Sante Fe, NM (1965).

Hall, K. "The Factors Which Play a Role in the Solution of Lead by Water," *Journal American Water Works Association,* 29:293 (1937).

Hallden, O. S., "Underground Natural Gas Storage (Herscher Dome)," Ground Water Contamination, U.S. Department of Health, Education and Welfare, R. A. Taft Sanitary Engineering Center, Technical Report W61-5, Cincinnati, Ohio, 218 pp. (1961).

Hammerton, C. "The Corrosion of Cement and Concrete," *Sewage Works Journal,* 17:403 (1945).

Hanes, R. E. et al. "Effects of Deicing Salts on Water Quality and Biota," Highway Research Board, National Cooperative Highway Research Program, Report 91, 70 pp. (1970).

Harmeson, R. H. and O. W. Vogel. "Artificial Recharge and Pollution of Ground Water," *Ground Water,* 1(1):11–15 (1963).

Haque, R. *Dynamic, Exposure and Hazard Assessment of Toxic Chemicals.* Ann Arbor, Mich.:Ann Arbor Science, Pub., Inc. (1980).

Harmon, B. "Contamination of Groundwater Resources," *Civil Engineering,* 11:343 (1941).

Harvey, E. J. and J. Skelton. "Hydrologic Study of a Waste-Disposal Problem in a Karst Area at Springfield, Missouri," Geological Survey Research 1968, U.S. Geological Survey Prof. Paper 600-C, pp. C217–C220 (1968).

Hassan, A. A. "Effects of Sanitary Landfills on Quality of Groundwater—General Background and Current Study," Paper presented at Los Angeles Forum on Solid Waste Management (1971).

Hedgepeth, J. W. "An Experiment in Limnological Method," *Journal Ent. Zool.,* 35:62 (1943); *Water Pollution Abs.,* 17 (July 1944).

Heller, V. G. "The Effect of Saline and Alkaline Waters on Domestic Animals," *Oklahoma A and M College Expt. Station Bull.,* No. 217 (Dec. 1933).

Hem, J. D. *Study and Interpretation of the Chemical Characteristics of Natural Water.* 1st editon, Geological Survey Water-Supply Paper 1473 (1959).

Hem, J. D. *Study and Interpretation of the Chemical Characteristics of Natural Water.* 2nd edition, U.S. Geological Survey Water-Supply Paper 1473, 363 pp. (1970).

Hemphill, D. D., ed. *Trace Substances in Environmental Health, Vol. VI.* University of Missouri (1972).

Henkel, H. O. "Surface and Underground Disposal of Chemical Wastes at Victoria, Texas," *Sewage and Industrial Wastes,* 25(9):1044–1049 (1953).

Hinman, J. J., Jr. "Desirable Characteristics of a Municipal Water Supply," *Journal American Water Works Association,* 30:438 (1938).

Hodgkiss, M. T. and J. W. Moodie. "Genomic Analysis of RNA Viruses Isolated From Water," *Water Science and Technology,* 17(10):16–22 (1984).

Hoehn, E. and R. H. Gunthen. "Distribution of Metal Pollution in Groundwater Determined from Sump Sludges in Wells," *Water Science and Technology,* 17(9):115–132 (1984).

Hoffman, E. J. et al. "Petroleum Hydrocarbons in Urban Runoff from a Commercial Land-Use Area," *Journal Water Pollution Control Federation,* 54:1517 (1982).

Holder, P. W. *Pesticides and Groundwater Quality.* Board of Agriculture, National Research Council, Washington, D.C.:National Academy Press (1986).

Holtzer, T. L. "Limits to Growth and Septic Tanks," in *Water Pollution Control in Low Density Areas.* W. J. Jewell and R. Swan, eds., Hanover, NH:University Press of New England, pp. 65–74 (1975).

Hubbert, M. K. and D. G. Willis. "Mechanics of Hydraulic Fracturing," *Journal of Petroleum Technology,* American Institute of Mining, Metallurgical Engineers Petroleum Division, Trans., T.P. 4597, pp. 153–168 (1957).

Huet, M. "pH Value and Reserves of Alkalinity," *Commun. Sta. Rech. Groenendoel, D.,* No. 1; *Water Pollution Abs.,* 21:254 (1948).

Hughes, G. M. and K. Cartwright. "Scientific and Administrative Criteria for Shallow Waste Disposal," *Civil Engineering,* 42(3):70–73 (1972).

Hughes, G. M. et al. "Hydrogeology of Solid Waste Disposal Sites in Northeastern Illinois," U.S. Environmental Protection Agency, Rep. SW-12D, 154 pp. (1971).

Huling, E. E. and T. C. Hollocher. "Groundwater Contamination by Road Salt: Steady-State Concentrations in East Central Massachusetts," *Science,* 176:288–290 (1972).

Hume, D. N. "Analysis of Water for Trace Metals," in *Equilibrium Concepts in Natural Water Systems,* Adv. Chem. Ser. NO. 67, American Chemical Society, Washington, D.C., pp. 30–44 (1967).

Hunter, J. V. et al. "Contribution of Urban Runoff to Hydrocarbon Pollution," *Journal Water Pollution Control Federation,* 51:2129 (1979).

Huyakorn, P. F. et al. "Saltwater Intrusion in Aquifers. Development and Testing of a Three-dimensional Finite Element Model," *Water Resources Research,* 23(2):293–312 (1987).

Iglehart, C. F. "North American Drilling Activity in 1971," *Bulletin, American Association of Petroleum Geologists,* 56(7):1145–1174 (1972).

Illvovitch, M. "World Water Resources Present and Future," *Ambio.,* 6:13–21 (1977).

Irwin, J. H. and R. B. Morton. "Hydrogeologic Information on the Glorieta Sandstone and the Ogallala Formation in the Oklahoma Panhandle and Adjoining Areas as Related to Underground Waste Disposal," U.S. Geological Survey Circular 630, 26 pp. (1969).

James, C. et al. "How Much Water in a 12-Ounce Can? A Perspective on Water-Use Information," U.S. Geological Annual Report, Fiscal Year 1976, Washington, D.C.:U.S. Govt. Printing Office (1980).

Javandel, I. C. et al. *Groundwater Transport: Handbook of Mathematical Models.* Water Resources Monograph, 10. I. Juandel et al., eds., AGU, Washington, D.C. (1984).

Jenkins, C. T. "Computation of Rate and Volume of Stream Depletion by Wells," *U.S. Geological Survey Techniques of Water Resources Investigations,* Bk. 4, Chap. D-1, Washington, D.C. (1970).

Jepsen, A. *Jord/Vand Hygiejne.* Copenhagen:A/S Carl Fr. Mortensen, 223 pp. (1972).

Johns, W. D. and W. H. Huang. "Distribution of Chlorine in Terrestrial Rocks," *Geochim. et Cosmochim. Acta.,* 31:35–40 (1967).

Johnson, W. R. et al. "Insecticides in Tile Drainage Effluent," *Water Resources Research,* 3(2):525–537 (1967).

Johnson Division. *Groundwater and Wells.* Saint Paul, Minn.:UOP Inc. Edward E. Johnson, Inc. (1975).

Jones, P. H. and E. Shuter. "Hydrology of Radioactive Waste Disposal in the MTR-ETR Area, National Reactor Testing Station, Idaho," Short Papers in Geology and Hydrology, U.S. Geologic Survey Prof. Paper 450-C, pp. C113–C116 (1962).

Jones, E. E., Jr. "Where Does Water Quality Improvement Begin?" *Ground Water,* 9(3):24–28 (1971).

Jorgensen, P. H. and E. Lund. "Detection and Stability of Enteric Viruses in Sludge, Soil, and Groundwater," *Water Science and Technology,* 17(10):185–196 (1984).

Jorgensen, D. G. "An Aquifer Test Used to Investigate a Quality of Water Anomaly," *Ground Water,* 9(3):24–28 (1968).

Josephson, J. "Groundwater Strategies," *Environmental Science and Technology,* 14(9):1031–1032 (1980).

Junge, E. *Air Chemistry and Radioactivity.* New York, London:Academic Press, 388 pp. (1963).

Kaiser, E. R. "Chemical Analysis of Refuse Components," *Proceedings 1966 Incineration Conference,* Am. Soc. of Mech. Engr., New York, New York (1966).

Kaufman, M. I. "Subsurface Waste Injection," *Journal Irrigation and Drainage Division,* American Soc. of Civil Engineers, 99(IR1):53–57 (1973).

Keeley, J. F. et al. "Evolving Concepts of Subsurface Contaminant Transport," *Journal Water Pollution Control Federation,* 58(5) (1986).

Kehoe, R. A. et al. "The Hygienic Significance of the Contamination of Water With Certain Mineral Constituents," *Journal American Water Works Association,* 36:645 (1944).

Kelly, W. P. et al. "Chemical Effects of Saline Irrigation on Soils," *Soil Science,* 49:95 (1939).

Keswick, B. H. et al. "Detection of Rotavirus in Treated Drinking Water," *Water Science and Technology,* 17(10):1–6 (1984).

Khan, I. A. "Determining Impact of Irrigation on Groundwater," *Journal Irrigation and Drainage Division,* Am. Soc. of Civil Engr., 106(IR4):331–334 (1980).

Kimmel, G. E. "Nitrogen Content of Ground Water in Kings County, Long Island, New York," U.S. Geological Survey Prof. Paper 800-D, pp. D199–D203 (1972).

Klock, J. W. "Survival of Coliform Bacteria in Wastewater Treatment Lagoons," *Journal of the Water Pollution Control Federation,* 43:2071–2083 (1971).

Knowles, D. B. "Hydrologic Aspects of the Disposal of Oil-Field Brines in Alabama," *Ground Water,* 3(2):22–27 (1965).

Kolbach, P. and G. Hausmann. "Influence of Sulphates and Chlorides of Alkaline Earths in the Brewing Liquor on the Composition of Wort," *Wochschr. Brau.,* 50:201 (1933); *Water Pollution Abstracts,* 7 (1934).

Kool, H. J. et al. "Toxicology Assessment of Organic Compounds in Drinking Water," *CRC Critical Reviews in Environmental Control, Vol. 12,* Issue 4 (1982).

Kovalev, M. M. "The Nature of Calculosis," *Klin. Med.* (U.S.S.R.) 31:10, 31 (1953): *Chemical Abstracts,* 48:12287 (1954).

Krauskopf, K. B. *Introduction to Geochemistry.* New York:McGraw-Hill Book Co., 721 pp. (1967).

Krone, R. B. et al. "Direct Recharge of Ground Water with Sewage Effluents," *Journal Sanitary Engineering Division, Vol. 83,* American Society of Civil Engineers, No. SA4, Paper 1335, 25 pp. (1957).

Kufs, C. et al. "Rating the Hazard Potential of Waste Disposal Facilities," *Proc. of the Nat. Conf. on Management of Uncontrolled Hazardous Wastes Sites,* Hazardous Materials Control Research Inst., Silver Spring, Maryland, pp. 30–41 (1980).

Lang, A. and H. Gruns. "On Pollution of Groundwater by Chemicals," *Gas u. Wasser,* 83(6):6; Abstract *Journal American Water Works Association,* 33:2075 (1940).

Langbein, W. B. "Salinity and Hydrology of Closed Lakes," U.S. Geological Survey Prof. Paper No. 412, 20 pp. Washington, D.C. (1961).

Lassen, J. "*Yersinia Enterocolitica* in Drinking Water," *Scandinavian Journal of Infectious Diseases,* 4:125–127 (1972).

Last, J. M. *Public Health and Preventive Medicine.* 11th edition, Maxcy-Rosenau, ed. Norwalk, Conn.:Appleton-Century Crofts (1980).

Laubusch, E. J. and C. S. McCammon. "Water as a Sodium Source and Its Relation to Sodium Restriction Therapy Patient Response," *Amer. Journ. of Public Health and the Nations' Health,* 45:10, 1337 (1955).

Law, J. P. "Irrigation Effects in Oklahoma and Texas," *Journal of Irrigation and Drainage Engineering,* 113(1) (Feb. 1987).

Leopold, L. B. "Hydrology for Urban Planning—A Guidebook on Hydrologic Effects of Urban Land Use," *U.S. Geological Survey Circular 554,* 18 pp. (1968).

Lewallen, M. J. "Pesticide Contamination of a Shallow Bored Well in the Southeastern Coastal Plains," *Ground Water,* 9(6):45–48 (1971).

Ley, J. B. "Deterioration of Concrete," *Water Pollution Abstracts,* 14 (Oct. 1941).

Lieber, M. and W. F. Welsh. "Contamination of Groundwater by Cadmium," *Journal American Water Works Association,* 46(6):541–547 (1954).

Lieber, M. et al. "Cadmium and Hexavalent Chromium in Nassau County Ground Water," *Journal American Water Works Association,* 56:739–747 (1964).

Little, A. D., Inc. "Salt, Safety, and Water Supply," Interim report of the special commission on salt contamination of water supplies and related matters, Commonwealth of Massachusetts, Senate No. 1485, 97 pp. (1973).

Ljunggren, P. "The Biogeochemistry of Manganese," *Geol. Forem. i Stockholm, Forh.,* 73:639–652 (1951).

Loch, J. P. G. and P. Lagas. "The Mobilization of Heavy Metals in River Sediment by Nitrilotriacetic Acid (NTA)," *Water Science and Technology,* 17(9):101–114 (1985).

Loehr, R. C. "Drainage and Pollution from Beef Cattle Feedlots," *Journal Sanitary Engineering Division, Vol. 96,* No. SA6, American Society of Civil Engineers, pp. 1295–1309, (1970).

Lorimer, J. C. et al. "Nitrate Concentrations in Groundwater Beneath a Beef Cattle Feedlot," *Water Resources Bulletin,* 8(5):999–1005 (1972).

Lowry, J. D. et al. "Point of Entry Removal of Radon from Drinking Water," *Journal American Water Works Association,* 79(4):162 (1987).

Lusczynski, H. J. and W. V. Swarzenski. "Salt-Water Intrusion in the United States," *Journal Hydraulics Division, Vol. 95,* No. HY5, American Soc. of Civil Engineers, pp. 1651–1669 (1966).

Lynn, R. D. and Z. E. Arlin. "Deep Well Construction for the Disposal of Uranium Mill Tailing Water by the Anaconda Company at Grants, New Mexico," *Soc. of Mining Engineers Trans.,* 223(3):230–237 (1962).

Maehler, C. Z. and A. E. Greenberg. "Identification of Petroleum Industry Wastes in Groundwaters," *Journal Water Pollution Control Federation,* 34(12):1262–1267 (1962).

Manufacturing Chemists Association. "Most Solid Wastes From Chemical Processes Used as Landfill on Company Property," *Currents* (1967).

Marsh, J. H. "Design of Waste Disposal Wells," *Ground Water,* 6(2):4–8 (1968).

Matis, J. R. "Petroleum Contamination of Ground Water in Maryland," *Ground Water,* 9(6):57–61 (1971).

Matthess, G. *The Properties of Groundwater.* New York:John Wiley and Sons, Inc. (1982).

Mattigod, S. V. et al. "Trace Metal Speciation in a Soil Profile Irrigated With Waste Waters," *Water Science and Technology,* 17(9):133–143 (1984).

Mazurek, M. A. and B. R. T. Simoneit. *CRC Critical Reviews in Environmental Control.* 16(1):1–140 (1986).

McClelland, N. I. "Individual On-Site Wastewater Systems," *Proc. of 3rd Natl. Conf. Ann Arbor Publishers,* Ann Arbor, Mich. (1977).

McCutcheon, T. P. et al. *General Chemistry.* 2nd ed., Von Nostrand Co. (1936).

McGauhey, P. H. and R. B. Krone. "Soil Mantle as a Wastewater Treatment

System," Final Report, SERL Report No. 67-11, Sanitary Engineering Research Laboratory, University of California, Berkeley (1967).

McKee, J. E. et al. "Gasoline in Groundwater," *Journal Water Pollution Control Federation*, 44:293–302 (1972).

McKee, J. E. and H. W. Wolf. *Water Quality Criteria*. State Water Quality Control Board, Publ. Sacramento, California, Pub. 3-A (1963).

McKee, J. E. and H. E. Wolf. *Water Quality Criteria*. California State Water Resources Board, Sacramento California, Publ. 3-A (1972).

McKinney, W. A. "Solution Mining," *Mining Engineering*, pp. 56–57 (1973).

McLean, D. D. "Subsurface Disposal: Precautionary Measures," *Industrial Water Engineering*, 6(8):20–22 (1969).

McMillion, L. G. "Hydrologic Aspects of Disposal of Oil-Field Brines in Texas," *Ground Water*, 3(4):36–42 (1965).

McMillion, L. G. "Ground Water Reclamation by Selective Pumping," *Trans. of AIME*, 250:11–15 (1971).

Meichtry, T. M. "Leachate Control System," Paper presented at Los Angeles Forum on Solid Waste Management (1971).

Meinzer, O. E. "The History and Development of Ground-Water Hydrology," *Journ. Wash. Acad. Sci.*, 24(1):6 (1934).

Merkel, R. H. "The Use of Resistivity Techniques to Delineate Acid Mine Drainage in Ground Water," *Ground Water*, 10(5):38–42 (1972).

Mertz, R. C. and R. Stone. "Progress Report on Study of Percolation through a Landfill," U.S. Public Health Service Research Grant SW 00028-07 (1967).

Miller, D. W. "Basic Elements of Ground Water Contamination," in *Seminar on the Fundamentals of Groundwater Quality Protection*, presented by Geraghty and Miller, Inc., and American Ecology Services, Inc., Cherry Hill, N.J. (1981).

Miller, W. D. "Subsurface Distribution of Nitrates Below Commercial Cattle Feedlots, Texas High Plains," *Water Resources Bulletin*, 7(5):941–950 (1971).

Minear, R. A. "Analytical Techniques for Measuring and Monitoring Trace Metals," *Journal American Water Works Association*, 67:9–14 (1975).

Mink, L. L. et al. "Effect of Early Day Mining Operations on Present Day Water Quality," *Ground Water*, 10(1):17 (1972).

Mink, J. F. "Excessive Irrigation and the Soils and Ground Water of Oahu, Hawaii," *Science*, 135(3504):672–673 (1962).

Monogham, G. W. and G. J. Larson. "A Computerized Ground-Water Resources Information System," *Ground Water*, 23:233–239 (1985).

Moore, G. T. et al. "Epidemic Giardiasis at a Ski Resort," *New England Journal of Med.*, 281:402–404 (1976).

Moorhouse, J. L. "Groundwater Protection: A Contrast in State Style," *Water/Engineering and Management*, pp. 23–25 (March 1985).

Moriarity, F. *Ecotoxicology: the Study of Pollutants in Ecosystems*. New York, London:Academic Press (1983).

Morris, D. A. et al. "Hydrology of Subsurface Waste Disposal," National Reactor Testing Station, Idaho—Annual Progress Report, U.S. Geological Survey Open-File Report (IDO-22046), 228 pp. (1964).

Morrisson, A. "If Your City's Well Water Has Chemical Pollutants, Then What?" *Civil Engineering*, 51(9):65–67 (1981).

Muhlberger, C. W. "Possible Hazards from Chemical Contamination in Water Supplies," *Journal American Water Works Association*, 42:1027 (1950).

Muss, D. L. "Relationship between Water Quality and Deaths from Cardiovascular Disease," *Journal American Water Works Associaton*, 54:1371–1378 (1962).

Myers, H. C. "Manganese Deposits in Western Reservoirs and Distribution Systems," *Journal American Water Works Association*, 53:579 (1961).

National Academy of Sciences. *Water Quality Criteria*. National Academy of Engineering, Washington, D.C.:U.S. Government Printing Office (1974).

National Academy of Sciences. *Drinking Water and Health*. (National Research Council) Nat. Acad. of Sci., Washington, D.C. (1977).

National Research Council. *Drinking Water and Health, Vol(s). 1–5*. Washington, D.C.:National Academy Press (1977–83).

National Research Council. *Groundwater Contamination. Studies in Geophysics*. Washington, D.C.:National Academy Press (1984).

National Research Council. *Pesticides and Groundwater Quality*, Board of Agriculture, Washington, D.C.:National Academy Press (1986a).

National Research Council. *Groundwater Quality Protection, State and Local Strategies*. Washington, D.C.:National Academy Press (1986b).

National Resources Defense Council, Inc. Comments of the Natural Resource Defense Council, Inc. on National Revised Primary Water Regulations. "Volatile Synthetic Organic Chemicals in Drinking Water," advanced notice of proposed rulemaking, *47 Federal Register*, 9350–9358 (Mar. 4, 1982) (Sept. 30, 1982).

Nebeker, R. L. and L. T. Lakey. "Liquid Waste Management at the NRTS Test Reactor Area," *Nuclear Engineering, Vol. 68*, Part 23, No. 123, pp. 10–17 (1972).

Neefe, J. R. and J. Stokes. "An Epidemic of Infectious Hepatitis Apparently Due to a Waterborne Agent," *Journal of the American Medical Association*, 128:1063–1075 (1945).

Negas, S. S. "The Physiological Aspects of Mineral Salts in Public Water Supplies," *Journal American Water Works Association*, 30:242 (1938).

New Mexico E. I. D. "Regulations Governing Water Supplies and Liquid Waste Disposal," New Mexico Health and Environment Department, Environmental Improvement Division, Santa Fe, NM (1985).

Newport, B. D. "Salt Water Intrusion in the United States," Rep. 600/8-77-011; U.S. Environmental Protection Agency, Washington, D.C. (1977).

Nie, N. H. *SPSS-X, User's Guide*. New York, N.Y.:McGraw-Hill Book Co. (1983).

Nightingale, H. I. and R. L. McCormick. "Chemical Quality of Perched Septic Tank Effluent for Plant Use and Recharge," *Journal Water Pollution Control Federation*, 57(9) (1985).

Nightingale, H. I. "Statistical Evaluation of Salinity and Nitrate Content and Trends Beneath Urban and Agricultural Areas—Fresno California," *Ground Water*, 3(1):22–29 (1970).

Norton, W. H. "Artesian Wells of Iowa," *Iowa Geological Survey*, 6:122–124 (1911).

Novac, J. T. et al. "Biodegradation of Methanol and Tertiary Butyl Alcohol in Subsurface Systems," *Water Science and Technology*, 17(9):71–86 (1984).

Oborn, E. T. "Intracellular and Extracellular Concentration of Manganese and Other Elements by Aquatic Microorganisms," U.S. Geological Survey Water-Supply Paper 1667-C, 18 pp. (1964).

Office of the Secretary of Transportation, Title 49—Transportation, *Federal Register, Vol. 35*, No. 5, Washington, D.C. (Jan. 1970).

Office of the Secretary of Transportation, *Federal Register, Vol. 36*, No. 86, Washington, D.C. (Apr. 1971).

Office of the Secretary of Transportation, *Federal Register, Vol. 37*, No. 180, Washington, D.C. (1972).

Office of Pipeline Safety. "Summary of Liquid Pipeline Accidents Reported on DOT form 7000-1 from January 1, 1971 through Dec. 31, 1971," Department of Transportation, Washington, D.C. (1972).

Office of Technology Assessment. "Protecting the Nation's Groundwater from Contamination," U.S. Congress, Office of Technology Assessment, OTA-0-233, Washington, D.C.:U.S. Government Printing Office (1984).

Olsthoorn, T. N. "The Power of the Electronic Worksheet: Modeling Without Special Programs," *Ground Water*, 34:381–390 (1985).

Ostroff, A. G. *Introduction to Oilfield Water Technology*. Englewood Cliffs, New Jersey:Prentice-Hall, Inc., 412 pp. (1965).

Owe, M. P. et al. "Contamination Levels in Precipitation and Urban Surface Runoff," *Water Resources Bulletin*, 18:863–868 (1982).

Parker, G. G. et al. "Water Resources of Southeastern Florida with Special Reference to Geology and Ground Water of the Miami Area," U.S. Geological Survey Water-Supply Paper 1255, 965 pp. (1955).

Parker, R. L. "Composition of the Earth's Crust," in *Data of Geochemistry*, 6th ed., U.S. Geological Survey Professional Paper 440-D, 19 pp. (1967).

Parks, W. W. "Decontamination of Groundwater at Indian Hill," *Journal American Water Works Association*, 51(5):644–646 (1959).

Patterson, R. J. et al. "Retardation of Toxic Chemicals in a Contaminated Outwash Aquifer," *Water Science and Technology*, 17(9):39–56 (1984).

Payne, R. D. "Saltwater Pollution Problems in Texas," *Journal Petroleum Technology*, 18:1401–1407 (1966).

Pelczar, M. J. and R. D. Reid. *Microbiology.* 2nd ed., New York:McGraw-Hill Book Company, 512 pp. (1965).

Pennypacker, S. P. et al. "Renovation of Wastewater Effluent by Irrigation of Forest Land," *Journal Water Pollution Control Federation,* 39(2):285–296 (1967).

Perkins, R. J. "Septic Tanks, Lot Size and Pollution of Water Table Aquifers," *Journal of Environmental Health,* 46(6):298–302 (1984).

Perlmutter, N. M. and A. A. Guerrera. "Detergents and Associated Contaminants in Ground Water at Three Public-Supply Well Fields in Southwestern Suffolk County, Long Island, New York," U.S. Geological Survey Water-Supply Paper 2001-B, 22 pp. (1970).

Perlmutter, N. M. and M. Lieber. "Dispersal of Plating Wastes and Sewage Contaminants in Ground Water and Surface Water, South Farmingdale-Massapequa Area, Nassau County, New York," U.S. Geological Survey Water-Supply Paper 1879-G, 67 pp. (1970).

Perlmutter, N. M. et al. "Movement of Waterborne Cadmium and Hexavalent Chromium Wastes in South Farmingdale, Nassau County, Long Island, New York," Short Papers in Geology and Hydrology, U.S. Geological Survey Professional Paper 475-C, pp. C179–C184 (1963).

Pernichele, A. D. "Geohydrology," *Mining Engineering,* pp. 67–68 (Feb. 1973).

Peters, J. A. and J. L. Rose. "Water Conservation by Reclamation and Recharge," *Journal Sanitary Engineering Division, Vol. 94,* American Society of Civil Engineers, No. SA4, pp 625–638 (1968).

Petri, L. R. "The Movement of Saline Ground Water in the Vicinity of Derby, Colorado," *Society for Water Treatment and Examination, Proc., Vol. 11,* Pt. 2, pp. 88–93 (1962).

Petroleum Equipment and Services. "Salt Water Disposal and Oilfield Water Conservation," *Petroleum Equipment and Services,* 30(4):22–28 (1967).

Pettit, B. M., Jr. and A. G. Winslow. "Geology and Ground-Water Resources of Galveston County, Texas," U.S. Geological Survey Water-Supply Paper 1416, 157 pp. (1957).

Pluhowski, E. J. "Urbanization and Its Effects on the Temperature of Streams and Long Island, New York," U.S. Geological Survey Professional Paper 627-D, 108 pp. (1970).

Popkin, R. A. and T. W. Bendixen. "Improved Subsurface Disposal," *Journal Water Pollution Control Federation,* 40(8):1499–1514 (1968).

Pozen, M. A. "Water in the Brewery," *Mod. Brew. Age,* 23(67) (1940); *Water Pollution Abstracts,* 14 (Oct. 1941).

Preul, H. C. "Contaminants in Groundwaters Near Waste Stabilization Ponds," *Journal Water Pollution Control Federation, Vol. 40,* No. 4, pp. 649–669 (1968).

Price, D. "Rate and Extent of Migration of a 'One-Shot' Contaminant in an Alluvial Aquifer in Keizer, Oregon," Geological Survey Research 1967, U.S. Geological Survey Professional Paper 575-B, pp. B217–B220 (1967).

Price, W. J. *Analytical Atomic Absorption Spectrometry.* London:Heyden and Son (1972).

Prichard, H. M. "The Transfer of Radon from Domestic Water to Indoor air," *Journal American Water Works Association,* 79(4):159 (1987).

Proctor, J. F. and I. W. Marine. "Geologic, Hydrologic, and Safety Considerations in the Storage of Radioactive Wastes in a Vault Excavated in Chrystalline Rock," *Nuclear Science Engineering,* 22(3):350–365 (1965).

Purtymun, W. D. et al. "Distribution of Radioactivity in the Alluvium of a Disposal Area at Los Alamos, New Mexico," Geological Survey Research 1966, U.S. Geological Survey Professional Paper 550-D, pp. D250–D252 (1966).

Pye, V. I. et al. *Groundwater Contamination in the United States.* Philadelphia, PA:University of Penn. Press (1983).

Qasim, S. R. and J. C. Burchinal. "Leaching from Simulated Landfills," *Journal Water Pollution Control Federation,* 42(3):371–379, Part 1 (1970).

Rail, C. D. "Groundwater Monitoring Within an Aquifer—A Protocol," *Journal of Environmental Health,* 48(3):128–132 (1985a).

Rail, C. D. *Plague Ecotoxicology: Including Historical Aspects of the Disease in the Americas and the Eastern Hemisphere.* Springfield, Ill.:Charles C. Thomas, Publ. (1985b).

Rail, C. D. *Summarization of Water Quality Data (Bernalillo County 1960–1976) the Inorganic Ions.* Environmental Services Division, Environmental Health Dept., City of Albuquerque, Volumes I and II, unpublished manuscripts, 700 pp. (1986).

Rankama, K. and T. G. Sahama. *Geochemistry.* 1st edition, Chicago, Ill.:Chicago University Press, (1950).

Rankama, K. and T. G. Sahama. *Geochemistry.* 2nd edition, Chicago, Ill.:Chicago University Press, 912 pp. (1960).

Ramade, F. *Ecotoxicology.* 2nd edition, Paris:Masson (1979).

Ramirez-Muniz, J. *Atomic-Absorption Spectroscopy and Analyses by Atomic Absorption Flame Photometry.* New York:Elsevier Publishing Co. (1968).

Rantz, S. E. "Urban Sprawl and Flooding in Southern California," U.S. Geological Survey Circular 601-B, 11 pp. (1970).

Raymond, J. R. and W. H. Bierschenk. "Hydrologic Investigations at Hanford," *Amer. Geophysical Union Trans.,* 38(5):724–729 (1957).

Reck, C. W. and E. J. Simmons. "Water Resources of the Buffalo-Niagara Falls Region," U.S. Geological Survey Circular 173, 26 pp. (1952).

Reichert, S. O. "Radionuclides in Groundwater at the Savannah River Plant Waste Disposal Facilities," *Journal Geophysical Research,* 67(11): 4363–4374 (1962).

Reichert, S. O. and J. W. Fenimore. "Lithology and Hydrology of Radioactive Waste-Disposal Sites, Savannah River Plant, South Carolina," *Geologic Soc. of Amer. Engineering Geology Case Histories,* No. 5, pp. 53–69 (1964).

Remson, I. et al. "Water Movement in an Unsaturated Sanitary Landfill," *Journal Sanitary Engineering Div., Vol. 94,* Amer. Soc. of Civil Engineers, No. SA 2, Paper 5904, pp. 307–317 (1968).

Remson, I. G. et al. *Numerical Methods in Subsurface Hydrology.* New York:Interscience (1971).

Resnik, A. V. and J. M. Rademacher. "Animal Waste Runoff—A Major Water Quality Challenge," *2nd Compendium of Animal Waste Management,* Paper No. 1, 21 pp. (June 1970).

Rice, I. M. "Salt Water Disposal in the Permian Basin," *Producers Monthly,* 32(3):28–30 (1968).

Richards, L. A. "Diagnosis and Improvement of Saline and Alkaline Soils," *U.S. Department of Agriculture Handbook, Vol. 60,* Washington, D.C., 160 pp. (1954).

Rinne, W. W. "Need for Saline Groundwater Data to Advance Desalting Technology," *Water Resources Research,* 6(5):1482–1486 (1970).

Robbins, J. W. and G. J. Kriz. "Relation of Agriculture to Groundwater Pollution: A Review," *Trans. Amer. Soc. of Agricultural Engineers,* 12(3):397–403 (1969).

Robbins, R. J. "Give Your Underground Storage Tank a Proper Burial," *Hazardous Materials and Waste Management Magazine* (Mar./Apr. 1987).

Robertson, J. B. and I. Kahn. "The Infiltration of Aldrin through Ottawa Sand Columns," *Geological Survey Research 1969,* "U.S. Geological Survey Professional Paper 650-C, pp. C219–C223 (1969).

Robinove, C. J. et al. "Saline-Water Resources of North Dakota," U.S. Geological Survey Water-Supply Paper 1428, 72 pp. (1958).

Robinson, J. W. *Atomic Absorption Spectroscopy.* New York:Marcel Dekker Co. (1975).

Roedder, E. "Problems in the Disposal of Acid Aluminum Nitrate High-Level Radioactive Waste Solutions by Injection into Deep Permeable Formations," U.S. Geological Survey Bulletin 1088, 65 pp. (1959).

Rold, J. W. "Pollution Problems in the Oil Patch," *Amer. Assoc. Petroleum Geologists Bulletin,* 55(6):807–809 (1971).

Rorabaugh, M. I. "Ground Water in Northeastern Louisville, Kentucky, with Reference to Induced Infiltration," U.S. Geological Survey Water-Supply Paper 1360-B, 168 pp. (1956).

Rose, J. L. "Injection of Treated Waste Water into Aquifers," *Water and Wastes Engineering,* 5(10):40–43 (1968).

Rose, J. L. "Advanced Waste Treatment in Nassau County, New York, Water Provided for Injection into Groundwater Aquifers," *Water and Wastes Engineering,* 7(2):38–39 (1970).

Rothschild, E. R. et al. "Investigation of Aldicarb in Groundwater in Selected Areas of the Central Sand Plain of Wisconsin," *Ground Water,* 20(4):437–445 (1982).

Rothstein, A. "Toxicology of the Minor Metals," University of Rochester, AEC Project, UR-262 (June 1953).

Russelman, H. B. and M. P. Turn. "Management of Septic Tank Solids," *Third Annual Illinois Private Sewage Disposal Symposium*, Toledo Area Council of Governments, Toledo, Ohio, pp 9–17 (1978).

Sabol, G. V. et al. "Irrigation Effects in Arizona and New Mexico," *Journal of Irrigation and Drainage Engineering*, 113(1) (Feb. 1987).

Safe Drinking Water Act. 42 United States Congress. 300f to 300j-9. Public Law No. 93-523 (1974).

Salvato, J. A. et al. "Sanitary Landfill—Leaching Prevention and Control," *Journal Water Pollution Control Federation*, 43(10):2084–2100 (1971).

Sartor, J. D. and G. B. Boyd. "Water Pollution Aspects of Street Surface Contaminants," URS Research Company. San Mateo, California; U.S. Environmental Protection Agency, Office of Research and Monitoring Report, R2-72-081, 236 pp. (1972).

Sattar, S. A. et al. "Rotavirus Survival in Raw and Treated Waters and Its Health Implications," *Water Science and Technology*, 17(10):1–6 (1984).

Savage, R. J. "Groundwater Protection: Working Without a Statute," *Journal Water Pollution Control Federation*, 58(5) (1986).

Sawyer, C. H. *Chemistry for Sanitary Engineers*. New York:McGraw-Hill Book Co., Inc. (1960).

Sayre, A. N. and V. T. Stringfield. "Artificial Recharge of Groundwater Reservoirs," *Journal of American Water Works Association*, 40:1152–1158 (1948).

Schaff, P. *A Dictionary of the Bible*. New York (1885).

Schmidt, K. D. and I. Sherman. "Effect of Irrigation on Groundwater Quality in California," *Journal of Irrigation and Drainage Engineering*, 113(1) (1987).

Schmidt, S. D. and D. R. Spencer. "The Magnitude of Improper Waste Discharges in an Urban Stormwater System," *Journal Water Pollution Control Federation*, 58(7) (1986).

Schmidt, D. E. "Nitrate in Groundwater of the Fresno-Clovis Metropolitan Area, California," *Ground Water*, 10(1):50–64 (1972).

Schneider, A. D. et al. "Movement and Recovery of Herbidices in the Ogallala Aquifer," *The Ogallala Aquifer—A Symposium*, Inter. Centr. for Arid and Semi-Arid Land Studies, Special Report No. 39, Texas Tech University, Lubbock, pp. 219–226 (1970).

Schneider, W. J. and A. M. Spieker. "Water for the Cities—The Outlook," U.S. Geological Survey Circular 601-A, 6 pp. (1969).

Schwarzenbach, R. P. and J. Westfall. "Sorption of Hydrophobic Trace Organic Compounds in Groundwater Systems," *Water Science and Technology*, 17(9):39–56 (1985).

Schweisfurth, R. "*Crenothrix Polyspora* Cohn as an Indicator of Organic Pollution of Groundwater," *Proc. Soc. Ecol.*, P. Muller and W. Junk, eds. Erlangen (1974).

Seaburn, G. E. "Effects on Urban Development on Direct Runoff to East Meadow Brook, Nassau County, Long Island, New York," U.S. Geological Survey Professional Paper 627-B, 14 pp. (1969).

Shirley, P. A. "Use of STORET as a Data Base for Ground-Water Quality Management," *Proc. Sixth National Ground Water Quality Symposium*, NWWA, Worthington, Ohio (1982).

Shohl, A. T. *Mineral Metabolism*. American Chemical Society Monograph No. 82, Reinhold Pub. Co. (1939).

Slade, J. S. "Viruses and Bacteria in a Chalk Well," *Water Science and Technology*, 17(10):111–126 (1984).

Slade, J. S. "Viruses and Drinking Water," *Journal Water Engineers and Scientists*, 39:71–81 (1985).

Slagle, K. A. and J. M. Stogner. "Oil Fields Yield New Deep-Well Disposal Technique," *Water and Sewage Works*, 116(6):238–244 (1969).

Slavin, W. *Atomic Absorption Spectroscopy*. New York:Wiley Interscience (1968).

Smith, H. F. "Subsurface Storage and Disposal in Illinois," *Ground Water*, 9(6):20–28 (1971).

Smith, H. V. "Potability of Water from the Standpoint of Fluorine Content," *American Journal of Public Health*, 25:434 (1935).

Smith, H. V. et al. "Mottled Enamel in the Salt Valley and the Fluorine Content of the Water Supplies," Arizona Agricultural Experiment Station Technical Bulletin 61 (1936); *Journal American Water Works Association*, 29:1201 (1937).

Smith, N. *Man and Water: A History of Hydro-Technology*. Great Britain: Charles Schribner's Sons (1975).

Sobsey, M. D. et al. "Detection of Hepatitis—A Virus (HAV) in Drinking Water," *Water Science and Technology*, 17(10):23–38 (1984).

Sollman, T. M. *A Manual of Pharmacology*. 8th ed., Phila., PA:W. B. Saunders, Co. (1957).

Sonneborn et al. "Health Effects of Inorganic Drinking Water Constituents, Including Hardness, Iodide, and Fluoride," *CRC Critical Reviews in Environmental Control, Vol. 13*, Issue 1 (1983).

Sopper, W. W. "Waste Water Renovation for Reuse: Key to Optimum Use of Water Resources," *Water Research*, 2(7):471–480 (1968).

Sposito, G. "Sorption of Trace Metals by Humic Materials in Soils and Natural Waters," *CRC Critical Reviews in Environmental Control, Vol. 16*, Issue 2 (1986).

Steel, R. G. and J. H. Torrie. *Principles and Procedures of Statistics*. New York, N.Y.:McGraw-Hill Book Co. (1960).

Stewart, B. A. "Nitrate and Other Pollutants under Fields and Feedlots," *Env. Sci. Tech.*, 1:736 (1967).

Stewart, B. A. et al. *Distribution of Nitrates and Other Water Pollutants under Fields and Corrals in the Middle South Platte Valley of Colorado*, U.S. Dept. of Agriculture, Agricultural Research Service, ARS 41-134, 206 pp. (1967).

Stewart, B. A. et al. "Agriculture's Effect on Nitrate Pollution of Groundwater," *Journal Soil and Water Conservation*, 23(1):13–15 (1968).

Stiff, M. J. "Copper/Bicarbonate Equilibria in Solutions of Bicarbonate Ion at Concentrations Similar to Those Found in Natural Water," *Water Research*, 5:171 (1971).

Stone, R. and W. F. Garber. "Sewage Reclamation by Spreading Basin Infiltration," *Proceedings American Society of Civil Engineers, Vol. 77*, No. 87, 20 pp. (1954).

Stumm, W. and J. J. Morgan. *Aquatic Chemistry*. New York:John Wiley and Sons, Inc. (1970).

Swenson, H. A. "The Montebello Incident," *Soc. for Water Treatment and Examination Proc., Vol. 11*, Pt. 2, pp. 84–88 (1962).

Swoboda, A. R. "Distribution of DDT and Toxaphene in Houston Black Clay on Three Watersheds," *Environmental Science and Technology*, 5:141–145 (1971).

SAS Institute, Inc. *SAS User's Guide*. Raleigh, N.C.:Spark Press (1979).

Talbot, J. S. "Some Basic Factors in the Consideration and Installation of Deep Well Disposal Systems," *Water and Sewage Works*, Reference No. 1968, pp. R213–R219 (1968).

Task Committee on Saltwater Intrusion. "Salt-Water Intrusion in the United States," *Journal Hydraulics Division, Vol. 95*, Amer. Soc. of Civil Engineers, No. HY5, pp. 1651–1669 (1969).

Taylor, C. R. H. *A Pacific Bibliography*. Printed matter related to the native peoples of Polynesia, Melanesia, and Micronesia. The Polynesian Society, Wellington, N.Z. (1951).

Taylor, F. B. et al. "The Case for Waterborne Infectious Hepatitis," *American Journal of Public Health*, 56:2093–2105 (1966).

Taylor, S. R. "Abundance of Chemical Elements in the Continental Crust—A New Table," *Geochim. et Cosmochim. Acta.*, 28:1273–1285 (1964).

Tchobanoglous, G. and R. Eliassen. "The Indirect Cycle of Water Reuse," *Water and Wastes Engineering*, 6(2):35–41 (1969).

Thomas, H. E. and W. J. Schneider. "Water as an Urban Resource and Nuisance," U.S. Geological Survey Circ. 601-D, 9 pp. (1970).

Thompson, D. R. "Complex Ground-Water and Mine-Drainage Problems from a Bituminous Coal Mine in Western Pennsylvania," *Bulletin Association of Engineering Geologists*, 9(4):335–346 (1972).

Todd, D. K. and C. F. Meyer. "Hydrology and Geology of the Honolulu Aquifer," *Journal of Hydraulics Division, Vol. 97*, American Society of Civil Engineers, No. HY2, pp. 233–256 (1971).

Todd, D. K. *Ground Water Hydrology*. New York:John Wiley & Sons (1959).

Todd, D. K. "Saltwater Intrusion of Coastal Aquifers in the United States,"

International Association of Scientific Hydrology, Publication No. 52, pp. 452–561 (1960).

Todd, D. K. *Groundwater Pollution in Europe—A Conference Summary,* GE73TMP-1, General Electric Co., Santa Barbara, Calif., 79 pp. (1973).

Tofflemire, T. J. and G. P. Brezner. "Deep Well Injection of Waste Water," *Journal Water Pollution Control Federation,* 43(7):1468–1479 (1971).

Tolman, C. F. *Ground Water.* New York, London:McGraw-Hill Book Co. (1937).

Toth, J. "Mapping and Interpretation of Field Phenomenon for Groundwater Reconnaissance in a Prairie Environment, Alberta, Canada," *Int. Assoc. Sci. Hydrol. Bull.,* 11(2):20–26 (1966).

Trelease, F. J. *Water Law: Resource Use and Environmental Protection."* St. Paul, Minn.:West Pub. Co. (1974).

Truhaut, R. "Ecotoxicology—A New Branch of Toxicology," in *Ecological Toxicology Research,* A. D. McIntyre and C. F. Mills, eds. *Proc. NATO Science Comm. Conf. Mt. Gabriel, Quebec, May 6–10, 1974.* 323 pp., New York:Plenum Press (1975).

Truhaut, R. "Ecotoxicology: Objectives, Principles, and Perspectives," *Ecotoxicology and Environmental Safety,* 1:151 (1977).

Tucker, W. E. "Subsurface Disposal of Liquid Industrial Wastes in Alabama—A Current Status Report," *Ground Water,* 9(6):10–19 (1971).

Tung, Y. K. and G. E. Koltermann. "Some Computational Experiences Using Embedded Techniques for Ground-Water Management," *Ground Water,* 23:455–464 (1985).

Turekian, K. K. and K. H. Wedepohl. "Distribution of the Elements in Some Major Units of the Earth's Crust," *Geol. Soc. America Bull.,* 72:175–192 (1961).

Tyler, C. "The Mineral Requirements and Metabolism of Poultry. III. Elements Other than Calcium and Phosphorus," *Nutritional Abs. and Rev.,* 19:263 (1949).

Ulrich, A. A. "Chloride Contamination of Ground Water in Ohio," *Journal American Water Works Association,* 47(2):151–152 (1955).

UNESCO. Aquifer Contaminaton and Protection: Studies and Reports in Hydrology. United Nations Educational, Scientific, and Cultural Organization, *Imprimevie de la Manutention,* Mayenne, France, 439 pp. (1980).

U.S. Department of Agriculture—Water. *The Yearbook of Agriculture,* Alfred Steffrud (ed.), Superintendent of Documents, Washington, D.C.:U.S. Govt. Printing Off. (1955).

U.S. Department of Health, Education, and Welfare, *Manual of Septic Tank Practice,* U.S. Dept. of HEW Public Health Service, Publication No. 526 (1967).

U.S. Department of the Interior. Geothermal Leasing Program. NTIS Accession No. PB 203 102-D, 156 pp., Washington, D.C. (1971).

U.S. Environmental Protection Agency. "Water Programs, Guidelines Establishing Test Procedures for Analysis of Pollutants," *Federal Register,* 38(199):28758–28760 (Oct. 16, 1973).

U.S. Environmental Protection Agency. *Polluted Groundwater: A Review of the Recent Literature.* Environmental Monitoring Series. EPA-600/4-74-001 (1974).

U.S. Environmental Protection Agency. *Quality Criteria for Water,* prepublication copy (1976).

U.S. Environmental Protection Agency. *Monitoring Groundwater Quality: Economic Framework and Principles,* Environmental Monitoring and Support Laboratory. Office of Res. and Dev. EPA-600/4-76/045, Las Vegas, NV (1976b).

U.S. Environmental Protection Agency. *Waste Disposal Practices and Their Effects on Groundwater.* Report to Congress, Washington, D.C., 512 pp. (1977).

U.S. Environmental Protection Agency. *Alternatives for Small Wastewater Treatment Systems. On-Site Disposal, Septage Treatment, and Disposal.* EPA Technology Transfer Seminar Publication, EPA-625/4-77-011 (1977b).

U.S. Environmental Protection Agency. *Water-Related Environmental Fate of 120 Priority Pollutants, Vols. 1 and 2.* EPA 440/4-79-029a (1979).

U.S. Environmental Protection Agency. *Groundwater Protection.* Washington, D.C., 36 pp. (1980).

U.S. Environmental Protection Agency. *Rapid Assessment of Potential Ground-Water Contamination Under Emergency Response Conditions.* Office of Health and Environmental Assessment, EPA-600/8-83-030, Washington, D.C. (1983).

U.S. Environmental Protection Agency, Committee on the Challenges of Modern Society (NATO/CCMS): Drinking Water Microbiology, *NATO/CCMS Drinking Water Series,* EPA 570/9-84-006 (1984a).

U.S. Environmental Protection Agency. *Ground-Water Protection Strategy.* Office of Ground Water Protection, Washington, D.C. (1984b).

U.S. Environmental Protection Agency. *Septage Treatment and Disposal Handbook.* EPA 625/6-84-009, Cinn., Ohio (1984c).

U.S. Environmental Protection Agency. *Practical Guide to Groundwater Sampling.* Robert S. Kerr Environmental Research Laboratory, EPA/600/2-85/104, Ada, Okla. (1985).

U.S. Environmental Protection Agency. *Pesticides in Ground Water: Background Document.* Office of Ground Water Protection (WH-550G), Washington, D.C. (1986).

U.S. Environmental Protection Agency. *Proposed Regulations for Underground Storage Tanks: What's in the Pipeline?* Office of Underground Storage Tanks, Washington, D.C. (1987).

U.S. General Accounting Office (U.S. GAO). *Ground Water Overdrafting Must be Controlled.* A Report to Congress by the U.S. Comptroller General, CED-80-86, 52 pp. (1980).

U.S. Geological Survey. *The Hydrologic Cycle.* Washington, D.C.:U.S. Government Printing Office (1978).

U.S. Geological Survey. *How Much Water in a 12-Ounce Can? A Perspective on Water-Use Information,* from U.S. Geological Survey Annual Report, Fiscal Year 1976, U.S. Government Printing Office (311-348/56) (1980).

U.S. Geological Survey Yearbook: 1980. *Ground-Water Contamination— No "Quick Fix" in Sight, Reprint from U.S. Geological Survey Yearbook.* Fiscal Year 1980, Washington, D.C.:U.S. Government Printing Press (1980).

U.S. Public Health Service. *Drinking Water Standards.* U.S. Public Health Service Pub. No. 956, Washington, D.C., 61 pp. (1962).

Valentine, R. L. et al. "Radium Removal Using Sorption to Filter Sand," *Journal American Water Works Association,* 79(4):170 (1987).

Valyashko, M. G. "Some General Rules with Respect to the Formation of the Chemical Composition of Natural Waters, Akad. Nauk SSSR Trudy Lab.," *Gidrogeol. Problem im: F.P. Savarenskogo,* 16:127–140 (1958).

Valyashko, M. G. "Geochemistry of Natural Waters," *Geokimiya,* pp. 1395–1407 (1967).

Van der Leeden, F. *Ground Water: A Selected Bibliography.* Port Washington, New York:Water Information Center, Inc. (1971).

Van der Warden et al. "Transport of Mineral Oil Components to Groundwater—1. Model Experiments on the Transfer of Hydrocarbons from a Residual Oil Zone to Trickling Water," *Water Research,* 5(5):213–226 (1971).

Varma, M. M. et al. "Trihalomethanes in Groundwater Systems," *Journal Environmental Systems,* 14(2), 1984–85 (1984).

Vaughn, J. C. et al. "Determination of Synthetic Detergent Content of Raw-Water Supplies," *Journal American Water Works Association,* 50:1343 (1958).

Vecchioli, J. and F. H. Ku. "Preliminary Results of Injecting Highly Treated Sewage Plant Effluent into a Deep Sand Aquifer at Bay Park, New York," U.S. Geological Survey Profession Paper 751A, 14 pp. (1972).

Veir, B. B. "Celanese Deep-Well Disposal Practices," *Water and Sewage Works,* 116(5):I/W 21–I/W 24 (1969).

Viets, F. G., Jr. "Water Quality in Relation to Farm Use of Fertilizer," *Bioscience,* 21(10):460–467 (1971).

Viraraghavan, T. and Hashem. "Trace Organics in Septic Tank Effluent," *Water, Air, and Soil Pollution,* 28:299–308 (1986).

Visher, F. N. and J. F. Mink. "Ground-Water Resources in Southern Oahu, Hawaii," U.S. Geological Survey Water-Supply Paper 1778, 133 pp. (1964).

Volker, H. et al. "Morphology and Ultrastructure of *Crenothrix polyspora* Cohn.," *Journal Bacteriology,* 131:306–313 (1977).

Walker, B. "The Present Role of the Local Health Departments in Environmental Toxicology," *Journal of Environmental Health,* 48(3):133–137 (1985).

Walker, T. R. "Ground-Water Contamination in the Rocky Mountain Arsenal Area, Denver, Colorado," *Geological Soc. of Amer. Bulletin,* 72(3):489–494 (1961).

Walker, W. R. and R. C. Stewart. "Deep-Well Disposal of Wastes," *Journal Sanitary Engineering Division, Vol. 94,* American Society of Civil Engineers, No. SA 5, Paper 6171, pp. 945–963 (1968).

Walton, G. "Survey of Literature Relating to Infant Methemoglobinemia Due to Nitrate-Contaminated Water," *American Journal of Public Health,* 41:986 (1951).

Waltz, J. P. "Methods of Geologic Evaluations of Pollution Potential at Mountain Homesites," *Ground Water,* 10(1):42–47 (1972).

Warner, D. L. "Deep-Well Injection of Liquid Waste," U.S. Public Health Service Environmental Health Service Publication, No. 999-WP-21, 55 pp. (1965).

Warner, D. L. "Deep-Well Waste Injection—Reaction with Aquifer Water," *Proc. American Society of Civil Engineers, Vol. 92,* No. SA 4, pp. 45–69 (1966).

Warner, D. L. "Deep-Wells for Industrial Waste Injection in the United States—Summary of Data," Federal Water Pollution Control Adminstration. Water Pollution Control Research Service Publication No. WP-20-10, 45 pp. (1967).

Warner, D. L. *Survey of Industrial Waste Injection Wells.* 3 Vols., Final Report, U.S. Geological Survey Contract No. 14-08-0001-12280, University of Missouri, Rolla, Missouri (1972).

Wei-chi Ying, et al. "Treatment of a Landfill Leachate in Powdered Activated Carbon Enhanced Sequencing Batch Bioreactors," *Env. Progress,* 6(1) (Feb. 1987).

Wells, H. G. *The Outline of History, Vol. 1.* The Review of Reviews Co., Macmillan Co. (1924).

Wen-Sen Chu et al. "An Evaluation of Data Requirements for Groundwater Contaminant Transport Modeling," *Water Resources Research,* 23(3):408–424 (1987).

Wesner, G. M. and D. C. Baier. "Injection of Reclaimed Wastewater into Confined Aquifers," *Journal American Water Works Association,* 62(3):203–210 (1970).

Westrick, J. J. et al. "The Ground Water Supply Survey: Summary of Volatile Organic Contaminant Occurrence Data," Technology Support Division, Office of Drinking Water, U.S. Environmental Protection Agency, Cinn., Ohio (1983).

White, D. E. and G. A. Waring. "Volcanic Emanations," in *Data of Geochemistry,* 6th edition, U.S. Geological Survey Professional Paper 440-K, pp. K1–K27 (1963).

White, D. E. et al. "Chemical Composition of Subsurface Waters," U.S. Geological Survey Professional Paper, No. 440-F, 67 pp. (1963).

White, D. E. "Magmetic, Connate, and Metamorphic Waters," *Geol. Soc. Am. Bull.,* 68:1659–1682 (1957).

White, D. E. "Summary of Chemical Characteristics of Some Waters of Deep Origin," in *Short Papers in the Geological Sciences,* U.S. Geological Survey Professional Paper 400-B, pp. B452–B454 (1960).

Wilcox, L. U. "The Quality of Water for Irrigation Use," *Tech. Bull. 962,* U.S. Department of Agriculture (Sept. 1948).

Williams, D. E. and D. G. Wilder. "Gasoline Pollution of a Groundwater Reservoir—A Case History," *Ground Water,* 9(6):50–56 (1971).

Wilmoth, B. M. "Occurrence of Salty Groundwater and Meteoric Flushing of Contaminated Aquifers," *Proc. of National Groundwater Quality Symposium,* U.S. Environmental Protection Agency Water Pollution Control Research Series 16060 GRB 08/71 (1971).

Wise, H. F. "Policy Implicatons of Urban Land Practices for Groundwater Quality," *Water and Sewage Works,* 84–85 (1977).

Witkowski, E. J. and J. F. Manneschmidt. "Ground Disposal of Liquid Wastes at Oak Ridge National Laboratory," *Proc. of 2nd Ground Disposal of Radioactive Wastes Conf.,* U.S. Atomic Energy Commission Div. of Technical Information Report TID-7628, Book 2, pp. 506–512 (1962).

Woodward, F. L. et al. "Experiences with Ground Water Contamination in Unsewered Areas in Minnesota," *American Journal of Public Health,* 51(8):1130–1136 (1961).

World Health Organization. "International Standards for Drinking Water," World Health Organization, Geneva (1958).

World Health Organization. "European Standards for Drinking Water," World Health Organization, Geneva (1961).

Wren, M. E. et al. "The Potential Toxicity of Uranium in Water," *Journal American Water Works Association,* 79(4):177 (1987).

Wulff. "The Qanats of Iran," *Scientific American,* 218:94–105 (1968).

Yosie, T. F. "EPA's Risk Assessment Culture," *Environmental Science and Technology,* 21(6):526–531 (1987).